30 Arduino™ Projects for the Evil Genius™

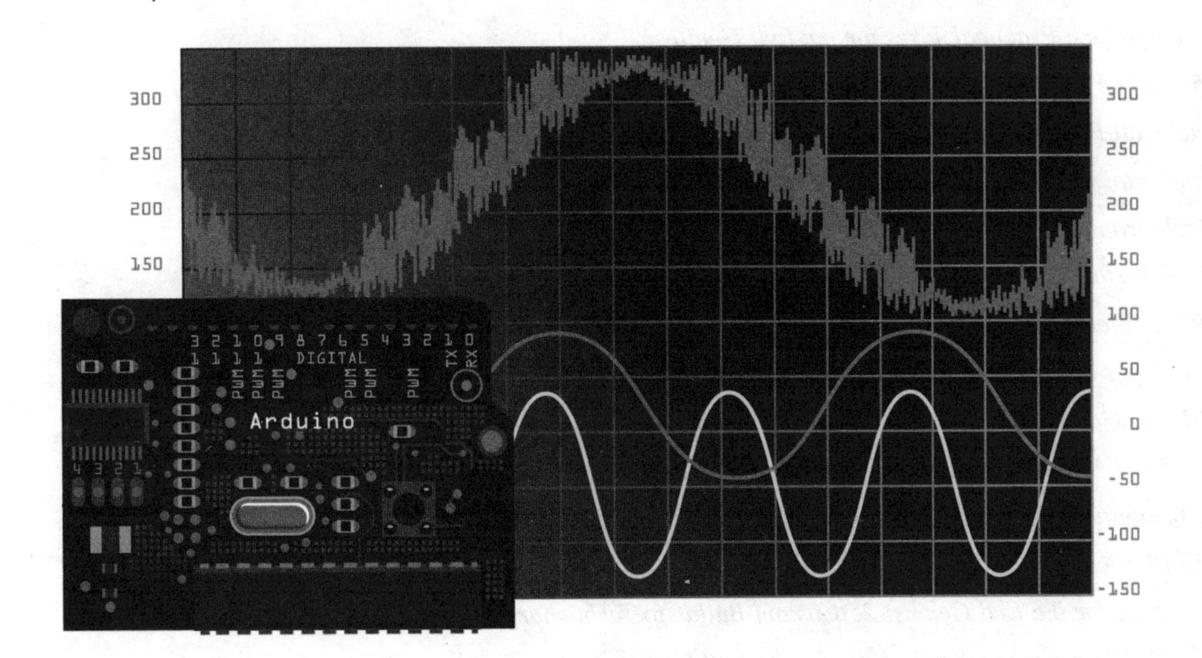

Evil Genius™ Series

Arduino + Android Projects for the Evil Genius
Bike, Scooter, and Chopper Projects for the Evil Genius
Bionics for the Evil Genius: 25 Build-It-Yourself Projects
Electronic Circuits for the Evil Genius, Second Edition: 64 Lessons with Projects
Electronic Gadgets for the Evil Genius: 28 Build-It-Yourself Projects
Electronic Sensors for the Evil Genius: 54 Electrifying Projects
15 Dangerously Mad Projects for the Evil Genius
50 Awesome Auto Projects for the Evil Genius
50 Green Projects for the Evil Genius
50 Model Rocket Projects for the Evil Genius
51 High-Tech Practical Jokes for the Evil Genius
46 Science Fair Projects for the Evil Genius
Fuel Cell Projects for the Evil Genius
Holography Projects for the Evil Genius
Mechatronics for the Evil Genius: 25 Build-It-Yourself Projects
Mind Performance Projects for the Evil Genius: 19 Brain-Bending Bio Hacks
MORE Electronic Gadgets for the Evil Genius: 40 NEW Build-It-Yourself Projects
101 Outer Space Projects for the Evil Genius
101 Spy Gadgets for the Evil Genius, Second Edition
123 PIC® Microcontroller Experiments for the Evil Genius
123 Robotics Experiments for the Evil Genius
125 Physics Projects for the Evil Genius
PC Mods for the Evil Genius: 25 Custom Builds to Turbocharge Your Computer
PICAXE Microcontroller Projects for the Evil Genius
Programming Video Games for the Evil Genius
Raspberry Pi Projects for the Evil Genius
Recycling Projects for the Evil Genius
Solar Energy Projects for the Evil Genius
Telephone Projects for the Evil Genius
30 Arduino Projects for the Evil Genius, Second Edition
tinyAVR Microcontroller Projects for the Evil Genius
22 Radio and Receiver Projects for the Evil Genius
25 Home Automation Projects for the Evil Genius

30 Arduino™ Projects for the Evil Genius™

Simon Monk

Second Edition

New York Chicago San Francisco Lisbon London Madrid
Mexico City Milan New Delhi San Juan Seoul
Singapore Sydney Toronto

McGraw-Hill Education books are available at special quantity discounts to use as premiums and sales promotions, or for use in corporate training programs. To contact a representative, please e-mail us at bulksales@mcgraw-hill.com.

30 Arduino™ Projects for the Evil Genius™, Second Edition

This book is printed on acid-free paper.

3 4 5 6 7 8 9 0 QVS QVS 20 19 18 17 16 15

ISBN 978-0-07-181772-1
MHID 0-07-181772-7

Sponsoring Editor
Roger Stewart

Editorial Supervisor
Stephen M. Smith

Production Supervisor
Pamela A. Pelton

Acquisitions Coordinator
Amy Stonebraker

Project Manager
Patricia Wallenburg, TypeWriting

Copy Editor
James Madru

Proofreader
Claire Splan

Indexer
Judy Davis

Art Director, Cover
Jeff Weeks

Composition
TypeWriting

To my late father, Hugh Monk, from whom I inherited a love for electronics.
He would have had so much fun with all this.

About the Author

Simon Monk has a bachelor's degree in cybernetics and computer science and a doctorate in software engineering. He has been an active electronics hobbyist since his school days and has written a number of books in the Evil Genius series, including *15 Dangerously Mad Projects for the Evil Genius* and *Arduino + Android Projects for the Evil Genius*. His other books include *Hacking Electronics* and *Programming Arduino: Getting Started with Sketches*.

Contents

Acknowledgments

I WOULD LIKE to thank my sons, Stephen and Matthew Monk, for their interest and encouragement in the writing of this book, their helpful suggestions, and their field testing of projects. Also, I could not have written this book without Linda's patience and support.

Finally, I would like to thank Roger Stewart and everyone at McGraw-Hill, who have been extremely supportive and enthusiastic and have been a pleasure to work with.

Introduction

Arduino interface boards provide the Evil Genius with a low-cost, easy-to-use technology to create their evil projects. A whole new breed of projects can now be built that can be controlled from a computer. Before long, the computer-controlled, servo-driven laser will be complete, and the world will be at the mercy of the Evil Genius!

This book will show the Evil Genius how to attach an Arduino board to their computer, to program it, and to connect all manner of electronics to it to create projects, including the computer-controlled, servo-driven laser mentioned earlier, a USB-controlled fan, a light harp, a USB temperature logger, a sound oscilloscope, and many more.

Full schematic and construction details are provided for every project, and most can be built without the need for soldering or special tools. However, the more advanced Evil Genius may wish to transfer the projects from a plug-in breadboard to something more permanent, and instructions for this are also provided.

So, What Is Arduino?

Well, Arduino is a small microcontroller board with a USB plug to connect to your computer and a number of connection sockets that can be wired up to external electronics, such as motors, relays, light sensors, laser diodes, loudspeakers, microphones, etc. Arduinos can be powered either through the USB connection from the computer or from a 9V battery. They can be controlled from the computer or programmed by the computer and then disconnected and allowed to work independently.

This book focuses on the most popular types of Arduino board, the Arduino Uno and the Arduino Leonardo.

At this point the Evil Genius might be wondering which top-secret government organization they need to break into in order to acquire an Arduino. Well, disappointingly, no evil deeds at all are required to obtain one of these devices. The Evil Genius needs to go no further than their favorite online auction site or search engine. Since the Arduino is an open-source hardware design, anyone is free to take the designs and create their own clones of the Arduino and sell them, so the market for the boards is competitive. An official Arduino costs about $30 and a clone often less than $20.

The software for programming your Arduino is easy to use and also freely available for Windows, Mac, and LINUX computers at no cost.

Arduino

Although Arduino is an open-source design for a microcontroller interface board, it is actually rather more than that, as it encompasses the software development tools that you need to program an Arduino board, as well as the board itself. There is a large community of construction, programming, electronics, and even art enthusiasts willing to share their expertise and experience on the Internet.

To begin using Arduino, first go to the Arduino site (www.arduino.cc) and download the software for Mac, PC, or LINUX. Chapter 1 provides step-by-step instructions for installing the software on all three platforms.

There are, in fact, several different designs of Arduino board. These are intended for different types of applications. They can all be programmed from the same Arduino development software, and in general, programs that work on one board will work on all.

In this book we use the Arduino Uno and Leonardo boards, apart from one project that uses the Arduino Lilypad. Nearly all the projects will work with both an Arduino Uno and an Arduino Leonardo, and many will also work with older Arduino boards such as the Duemilanove.

When you are making a project with an Arduino, you will need to download programs onto the board using a USB lead between your computer and the Arduino. This is one of the most convenient things about using an Arduino. Many microcontroller boards use separate programming hardware to get programs into the microcontroller. With Arduino, it's all contained on the board itself. This also has the advantage that you can use the USB connection to pass data back and forth between an Arduino board and your computer. For instance, you could connect a temperature sensor to the Arduino and have it repeatedly tell your computer the temperature.

You can either let your computer power the Arduino board through your USB cable or supply external power using a direct-current (DC) adapter. The power supply can be any voltage between 7 and 12 volts. So a small 9V battery will work just fine for portable applications. Typically, while you are making your project, you will probably power it from USB for convenience. When you are ready to cut the umbilical cord (disconnect the USB lead), you will want to power the board independently. This may be with an external power adaptor or simply with a 9V battery connected to a plug to fit the power socket.

There are two rows of connectors on the edges of the board. The row at the top of the diagram is mostly digital (on/off) pins, although some can be used as analog outputs. The bottom row of connectors has useful power connections on the left and analog inputs on the right.

These connectors are arranged like this so that so-called shield boards can be plugged onto the main board in a piggyback fashion. It is possible to buy ready-made shields for many different purposes, including

- Connection to Ethernet networks
- LCD displays and touch screens
- WiFi
- Sound
- Motor control
- GPS tracking
- And many more

You can also use prototyping shields to create your own shield designs. We will use these protoshields in one of our projects. Shields usually have through connectors on their pins, which means that you can stack them on top of each other. So a design might have three layers: an Arduino board on the bottom, a GPS shield on it, and then an LCD display shield on top of that.

The Projects

The projects in this book are quite diverse. We begin with some simple examples using standard LEDs and also the ultra-high-brightness Luxeon LEDs.

In Chapter 5 we look at various sensor projects for logging temperature and measuring light and pressure. The USB connection to the Arduino makes it possible to take the sensor readings in these projects and pass them back to the computer, where they can be imported into a spreadsheet and charts drawn.

We then look at projects using various types of display technology, including an alphanumeric LCD message board (again using USB to get messages from your computer), as well as seven-segment and multicolor LEDs.

Chapter 7 contains four projects that use sound as well as a simple oscilloscope. We have a simple project to play tunes from a loudspeaker and build up to a light harp that changes the pitch and volume of the sound by waving your hand over light sensors. This produces an effect rather like the famous Theremin synthesizer. The final project in this chapter uses sound input from a microphone. It is a VU meter that displays the intensity of the sound on an LED display.

Chapter 10 uses the special USB keyboard and mouse impersonation feature exclusive to the Arduino Leonardo in some interesting projects.

The final chapters contain a mixture of projects. Among others, there is, as we have already mentioned, an unfathomable binary clock using an Arduino Lilypad board that indicates the time in an obscure binary manner only readable by an Evil Genius, a lie detector, a motor-controlled swirling hypnotizer disk, and, of course, the computer-controlled, servo-guided laser.

Most of the projects in this book can be constructed without the need for soldering; instead, we use a breadboard. A breadboard is a plastic block with holes in it with sprung metal connections behind. Electronic components are pushed through the holes at the front. These are not expensive, and a suitable breadboard is also listed in the Appendix. However, if you wish to make your designs more permanent, the book shows you how to do that, too, using the prototyping board.

Sources for all the components are listed in the Appendix, along with some useful suppliers. The only things you will need in addition to these components are an Arduino board, a computer, some wire, and a piece of breadboard. The software for all the projects is available for download from www.arduinoevilgenius.com.

Without Further Ado

The Evil Genius is not noted for their patience, so in Chapter 1 we will show you how to get started with Arduino as quickly as possible. This chapter contains all the instructions for installing the software and programming your Arduino board, including downloading the software for the projects, so you will need to read it before you embark on your projects.

In Chapter 2 we take a look at some of the essential theory that will help you build the projects described in this book and go on to design projects of your own. Most of the theory is contained in this chapter, so if you are the kind of Evil Genius who prefers to just make the projects and find out how they work afterwards, you may prefer, after reading Chapter 1, to just to pick a project and start building. Then, if you get stuck, you can use the index or read some of the early chapters.

30 Arduino™ Projects for the Evil Genius™

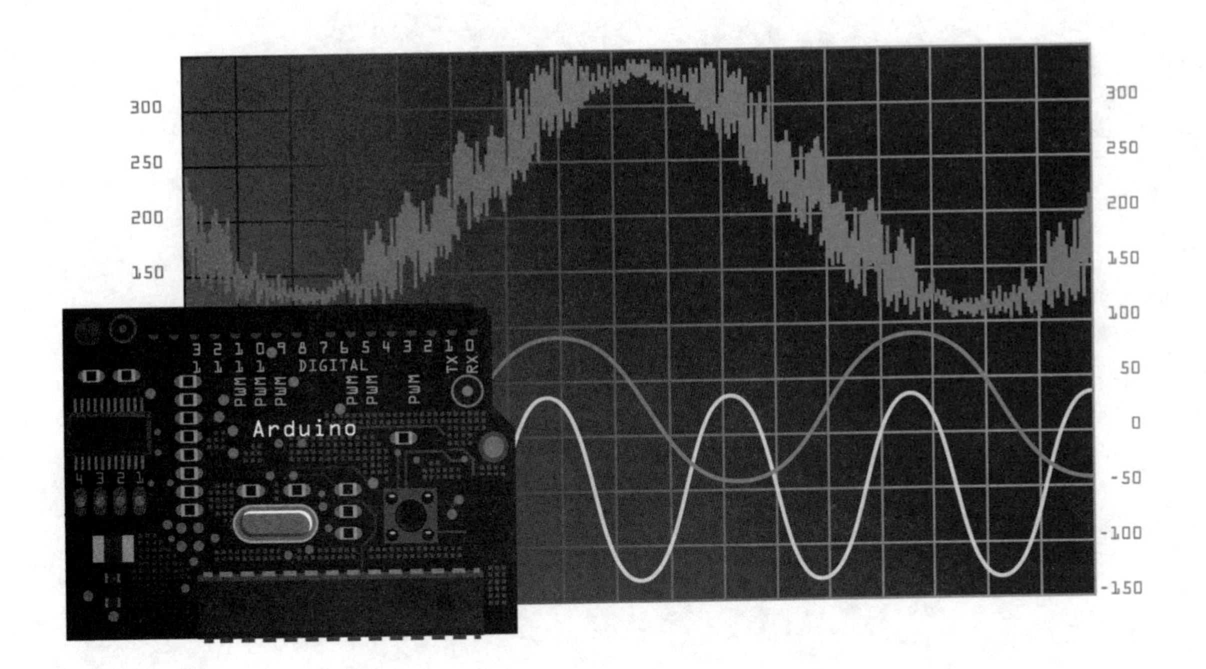

CHAPTER 1

Quickstart

THIS IS A CHAPTER for the impatient Evil Genius. Your new Arduino board has arrived, and you are eager to have it do something.

So, without further ado. . .

Powering Up

When you buy an Arduino Uno or Leonardo board, it is usually preinstalled with a sample Blink program that will make the little built-in LED flash. Figure 1-1 shows a pair of Arduino boards.

The light-emitting diode (LED) marked L is wired up to one of the digital input-output sockets on the board. It is connected to digital pin 13. This really limits pin 13 to being used as an output, but the LED only uses a small amount of current, so you can still connect other things to that connector.

All you need to do to get your Arduino up and running is supply it with some power. The easiest way to do this is to plug in it into the Universal Serial Bus (USB) port on your computer. For an Arduino Uno, you will need a type A-to-type B USB lead. This is the same type of lead that is normally used to connect a computer to a printer. For a Leonardo, you will need a micro-USB connector. You may get some messages from your operating system about finding new devices or hardware. Ignore these for now.

If everything is working okay, the LED should blink once every two seconds. The reason that new Arduino boards have this Blink sketch already installed is to verify that the board works. Clicking the Reset button should cause the LED to flicker momentarily. If this is the case, but the LED does not flash, then it may just be that the board has not been programmed with the Blink sketch; but do not despair, as once everything is installed, we are going to modify and install that script anyway as our first project.

Installing the Software

Now that we have our Arduino working, let's get the software installed so that we can alter the Blink program and send it down to the board. The exact procedure depends on what operating system you use on your computer. But the basic principle is the same for all.

Install the Arduino development environment, which is the program that you run on your computer that enables you to write sketches and download them to the Arduino board.

Figure 1-1 Arduinos Uno and Leonardo.

Install the USB driver that allows the computer to talk to the Arduino's USB port. It uses this for programming and sending messages.

The Arduino website (www.arduino.cc) contains the latest version of the software. In this book we have used Arduino 1.0.2.

Installation on Windows

The instructions that follow are for installing on Windows 7. The approach is much the same for Vista and XP. The only part that can be a little difficult is installing the drivers.

Follow the download link on the Arduino home page (www.arduino.cc), and select the download for Windows. This will start the download of the Zip archive containing the Arduino software, as shown in Figure 1-2. You may well be downloading a more recent version of the software than the version 1.0.2 shown (misleadingly, the Arduino team have not got around to renaming the Zip file yet).

The Arduino software does not distinguish between different versions of Windows. The

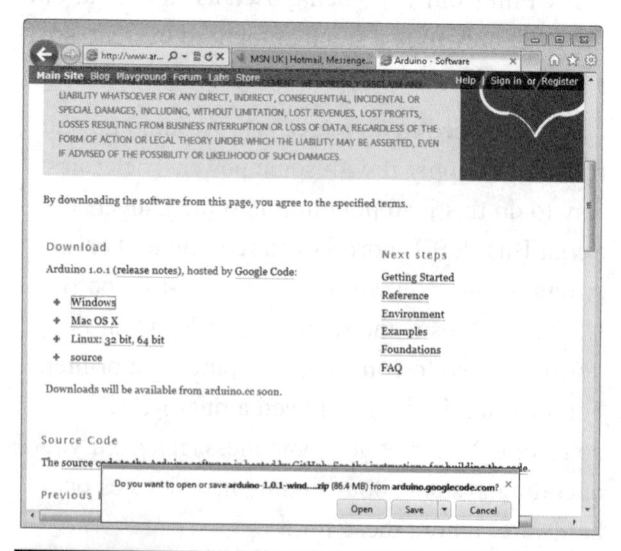

Figure 1-2 Downloading the Arduino software for Windows.

download should work for all versions, from Windows XP onward. The following instructions are for Windows 7.

Select the Save option from the dialog, and save the Zip file onto your desktop. The folder contained in the Zip file will become your main Arduino directory, so now unzip it onto your Desktop. You can move it somewhere else later if you wish.

You can do this in Windows by right-clicking the Zip file to show the menu in Figure 1-3 and selecting the Extract All option. This will open the Extraction Wizard, shown in Figure 1-4.

Extract the files to your Desktop.

This will create a new directory for this version of Arduino (in this case, 1.0.2) on your Desktop. You can, if you wish, have multiple versions of Arduino installed at the same time, each in its own folder. Updates of Arduino are fairly infrequent and historically have always kept pretty good compatibility with earlier versions of the software. So unless there is a new feature of the software that you want to use, or you have been having problems, it is by no means essential to keep up with the latest version.

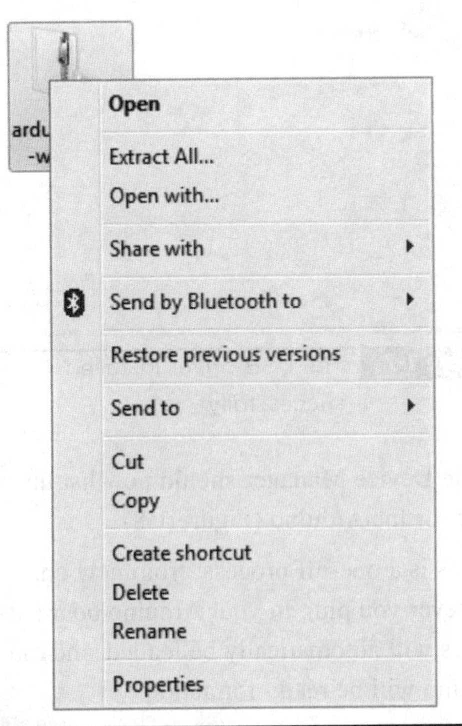

Figure 1-3 The Extract All menu option in Windows.

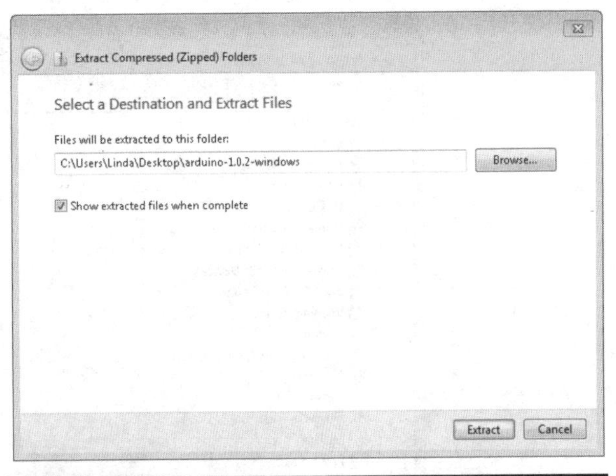

Figure 1-4 Extracting the Arduino file in Windows.

Now that we have got the Arduino folder in the right place, we need to install the USB drivers. If you have not already done so, plug your Leonardo or Uno into your computer. Depending on your version of Windows, there may be some half-hearted attempt by the operating system to install drivers. Just cancel this at the earliest opportunity; it is unlikely to work. Instead, you need to open the Device Manager. This is accessed in different ways depending on your version of Windows. In Windows 7, you first have to open the Control Panel, then select the option to view Icons, and you should find the Device Manager in the list.

Under the section "Other Devices," you should see an icon for "Unknown Device" with a little yellow warning triangle next to it. This is your Arduino (Figure 1-5).

Right-click on the "Unknown Device" and select the option "Update Driver Software." You will then be prompted to either "Search automatically for updated driver software" or "Browse my computer for driver software." Select the option to browse and navigate to the "arduino-1.0.2-windows\arduino1.0.2\drivers" (Figure 1-6).

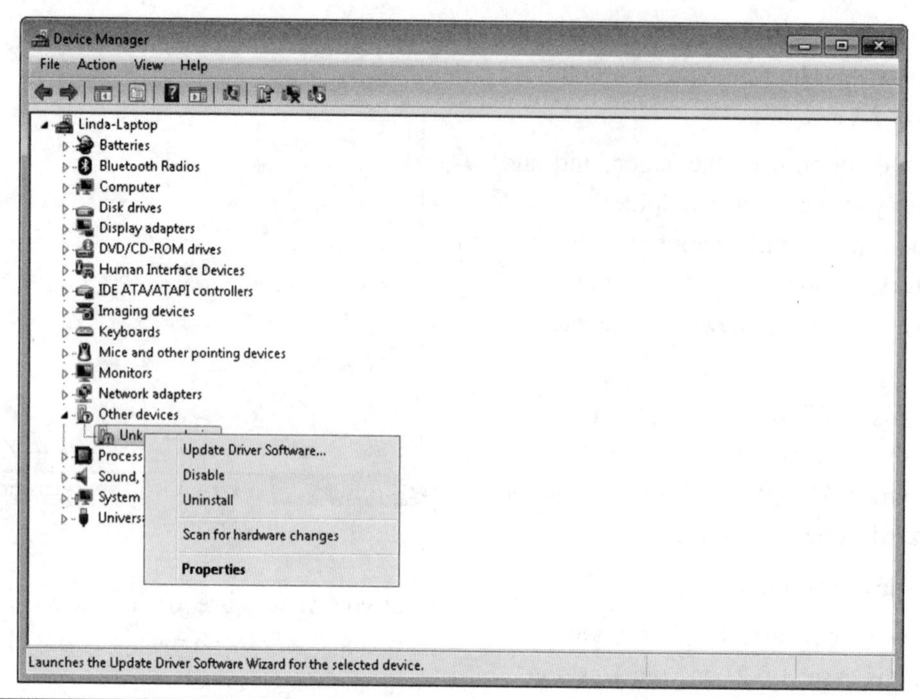

Figure 1-5 The Windows Device Manager.

Change the version numbers if you are using a different version of Arduino.

Click "Next," and you may get a security warning; if so, allow the software to be installed. Once the software has been installed, you will get a confirmation message like the one in Figure 1-7. Although the message will be different for a Leonardo, the procedure is identical.

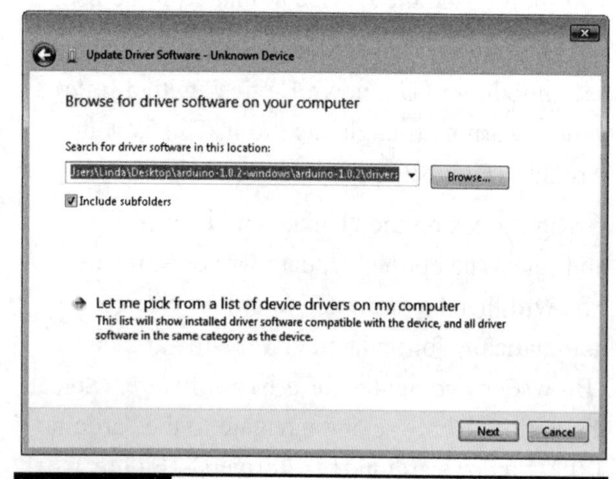

Figure 1-6 Browsing for the USB drivers.

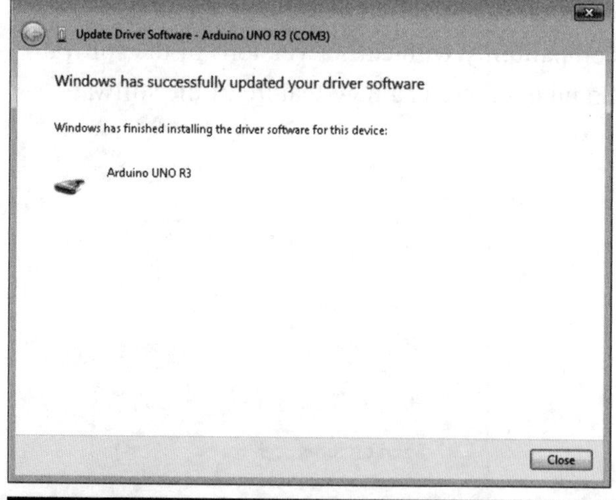

Figure 1-7 The USB driver Installed successfully.

The Device Manager should now list the right name for the Arduino (Figure 1-8).

This is a one-off process; from now on, whenever you plug in your Arduino board, its USB drivers will automatically be loaded, and the Arduino will be ready for action.

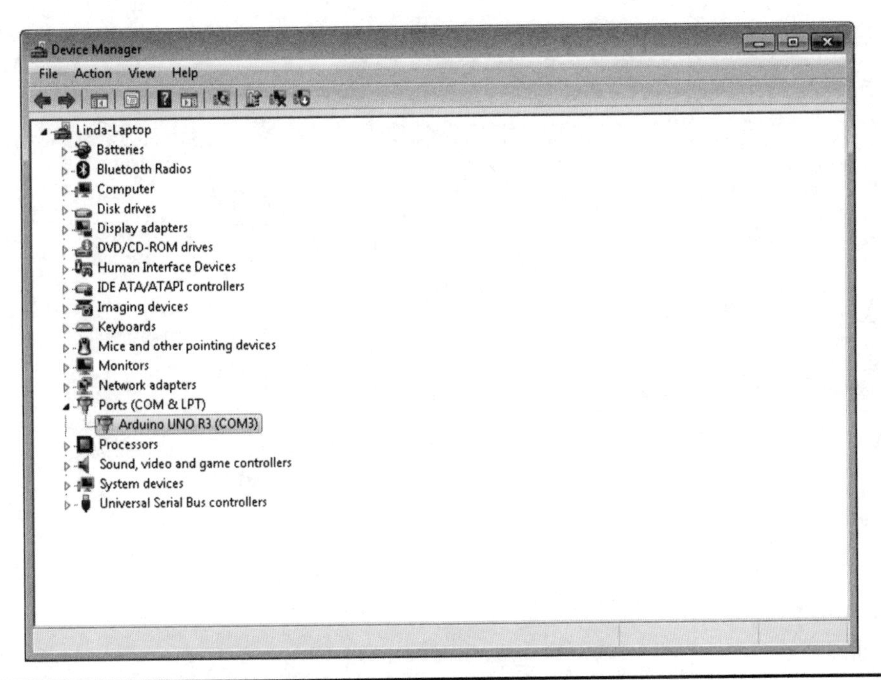

Figure 1-8 The Device Manager showing the Arduino.

Installation on Mac OS X

The process for installing the Arduino software on the Mac is a lot easier than on the PC.

As before, the first step is to download the file. In the case of the Mac, it is a Zip file. Once downloaded, double-click on the Zip file, which will extract a single file called "Arduino.app." This is the whole Arduino application; just drag it into your Applications folder.

You can now find and launch the Arduino software in your Applications folder. As you are going to use it frequently, you may wish to right-click its icon in the dock and set it to "Keep in Dock."

Installation on LINUX

There are many different LINUX distributions, and the instructions for each distribution are a little different. The Arduino community has done a great job of putting together sets of instructions for each distribution. So follow the link below and select one of the ten (at the time of writing) distributions on offer.

Configuring Your Arduino Environment

Whatever type of computer you use, you should now have the Arduino software installed on it. You now need to make a few settings. You need to specify the serial port that is connected to the Arduino board, and we need to specify the type of Arduino board that we are using. But first, you need to connect your Arduino to your computer using the USB lead or you will not be able to select the serial port.

Next, start the Arduino software. In Windows, this means opening the "Arduino" folder and clicking on the "Arduino" icon (selected in Figure 1-9). You may prefer to make a shortcut for the Desktop.

The serial port is set from the Tools menu, as shown in Figure 1-10 for the Mac and in Figure 1-11 for Windows 7—the list of ports for LINUX is similar to the Mac.

If you use many USB or Bluetooth devices with your Mac, you are likely to have quite a few options in this list. Select the item in the list that begins with "dev/tty.usbserial."

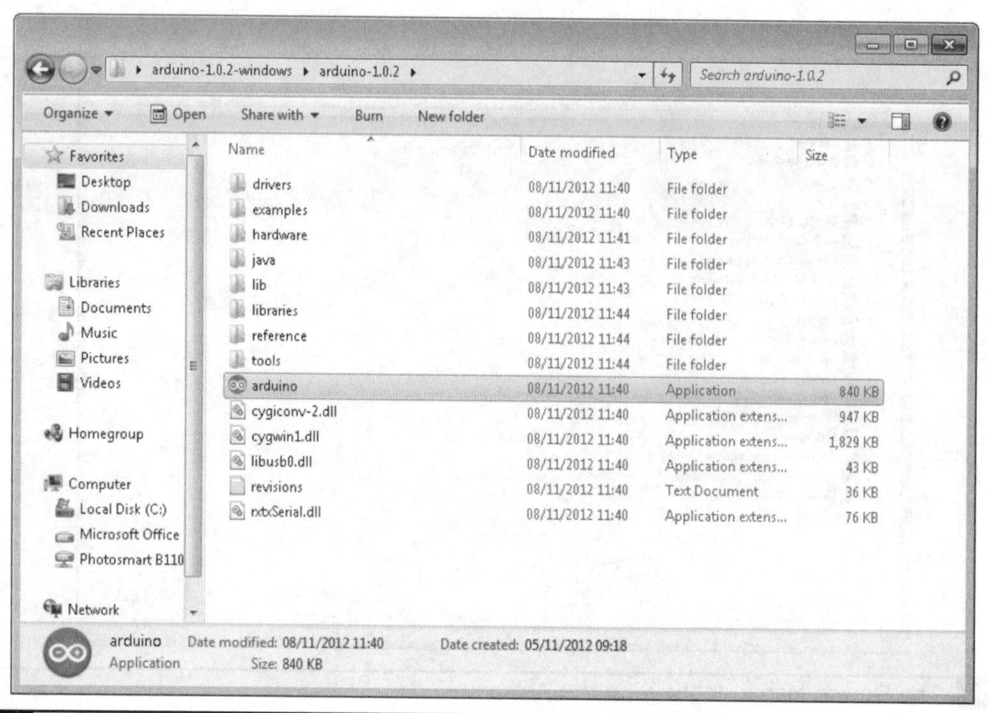

Figure 1-9 Starting Arduino in Windows.

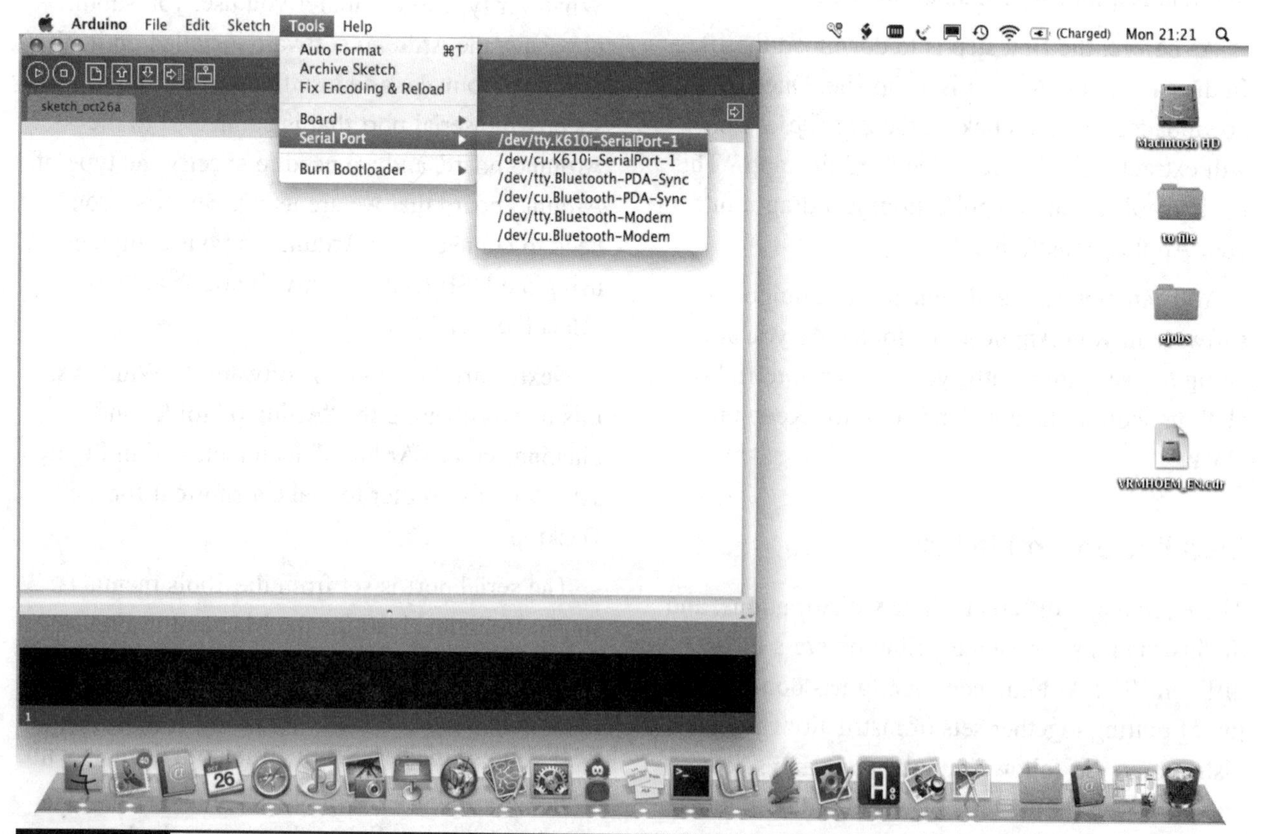

Figure 1-10 Setting the serial port on the Mac.

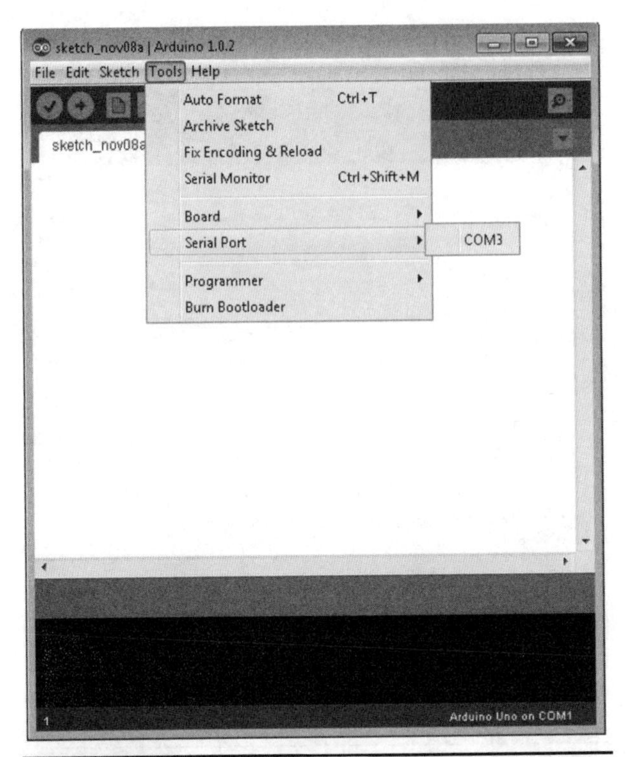

Figure 1-11 Setting the serial port on Windows.

On Windows, the serial port can just be set to COM3 or COM4, whichever shows up.

From the Tools menu, we can now select the board that we are going to use, as shown in Figure 1-12.

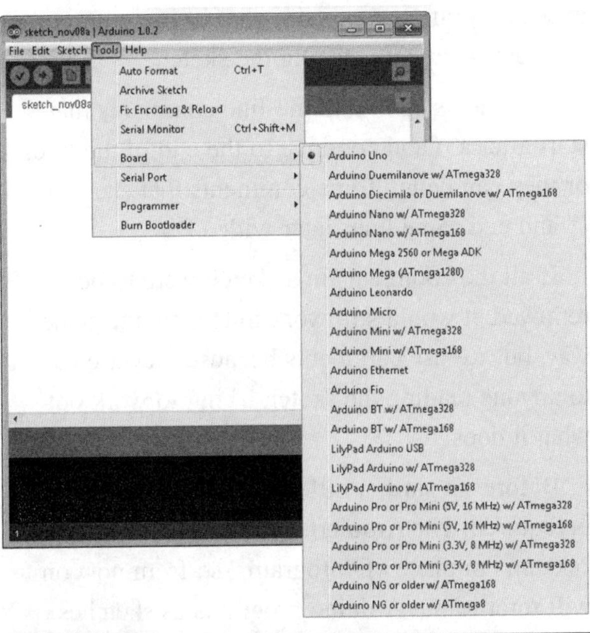

Figure 1-12 Setting the board.

Downloading the Project Software

The sketches (as programs are called in the Arduino world) used in the book are available as a single Zip file download. The whole download is less than a megabyte, so it makes sense to download the software for all of the projects, even if you only intend to use a few. To download them, browse to www.arduinoevilgenius.com and follow the download links for the second edition of this book.

Whatever your platform, the Arduino software expects to find all your sketches in your "Documents" folder, contained within a folder called "Arduino," which the Arduino software will create the first time it is run. So place the contents of the Zip file into that folder.

Note that each sketch comes in its own folders, and the sketches are numbered by project.

Project 1
Flashing LED

Having assumed that we have successfully installed the software, we can now start on our first exciting project. Actually, it's not that exciting, but we need to start somewhere, and this will ensure that we have everything set up correctly to use our Arduino board.

We are going to modify the example Blink sketch that comes with Arduino. We will increase the frequency of the blinking and then install the modified sketch on our Arduino board. Rather than blink slowly, our board will flash its LED quickly. We will then take the project a stage further by using a bigger external LED and resistor rather than the tiny built-in LED.

COMPONENTS AND EQUIPMENT

	Description	Appendix
	Arduino Uno or Leonardo	m1/m2
D1	5-mm red LED	s1
R1	270 Ω, 0.25 W resistor	r3
	Solderless beadboard	h1
	Jumper wires	h2

- In actual fact, almost any commonly available LED and 270 Ω resistor will be fine.
- The number in the Appendix column refers to the component listing in the Appendix, which lists part numbers for various suppliers.

Software

First, we need to load the Blink sketch into the Arduino software. The Blink sketch is included as an example when you install the Arduino environment. So we can load it using the File menu, as shown in Figure 1-13.

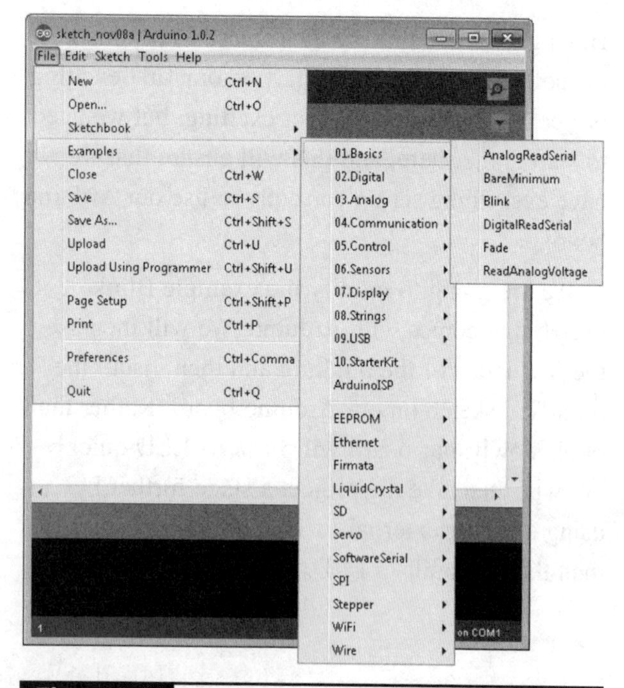

Figure 1-13 Loading the example Blink sketch.

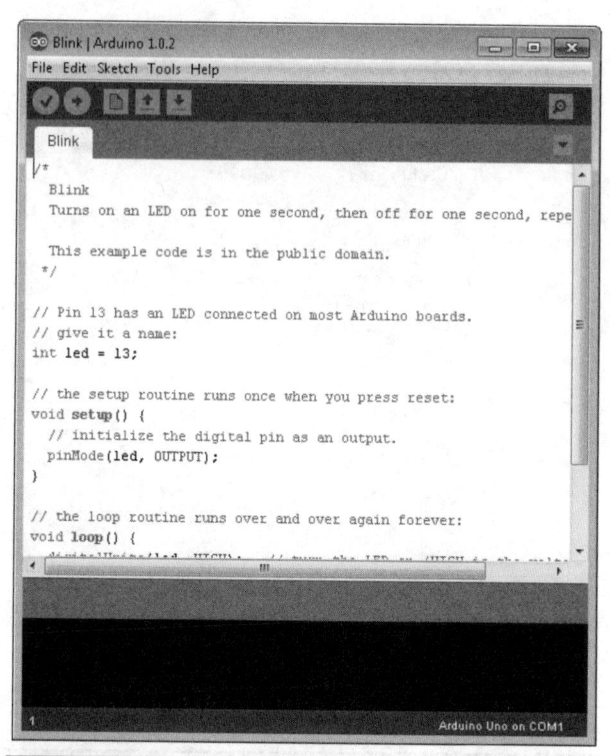

Figure 1-14 The Blink sketch.

The sketch will open in a separate window (Figure 1-14); you can, if you like, now close the empty window that opened when Arduino started.

The majority of the text in this sketch is in the form of comments. Comments are not actually part of the program but explain what is going on in the program to anyone reading the sketch.

Comments can be single-line comments that start after a // and continue to the end of the line, or they can be multiline comments that start with a /* and end some lines later with a */.

If all the comments in a sketch were to be removed, it would still work in exactly the same way, but we use comments because they are useful to anyone reading the sketch trying to work out what it does.

Before we start, a little word about vocabulary is required. The Arduino community uses the word "sketch" in place of "program," so from now on I will refer to our Arduino programs as sketches. Occasionally I may refer to "code." Code is programmer speak for a section of a program or even a generic term for what is written when creating a program. So someone might say, "I wrote a program to do that," or they could say, "I wrote some code to do that."

To modify the rate at which the LED will blink, we need to change the value of the delay so in the two places in the sketch where we have

```
delay(1000);
```

change the value in the parentheses to 200 so that it appears as

```
delay(200);
```

This is changing the delay between turning the LED on and off from 1000 milliseconds (1 second) to 200 milliseconds (1/5th of a second). In Chapter 3 we will explore this sketch further, but for now, we will just change the delay and download the sketch to the Arduino board.

With the board connected to your computer, click the Upload button on the Arduino. This is shown in Figure 1-15. If everything is okay, there will be a short pause, and then the two red LEDs on the board will start flashing away furiously as the sketch is uploaded onto the board. This should take around 5 to 10 seconds.

If this does not happen, check the serial port and board type settings as described in the previous sections.

When the completed sketch has been installed, the board will automatically reset, and if everything has worked, you will see the LED for digital port 13 start to flash much more quickly than before.

Hardware

At the moment, this doesn't really seem like real electronics because the hardware is all contained on the Arduino board. In this section we will add an external LED to the board.

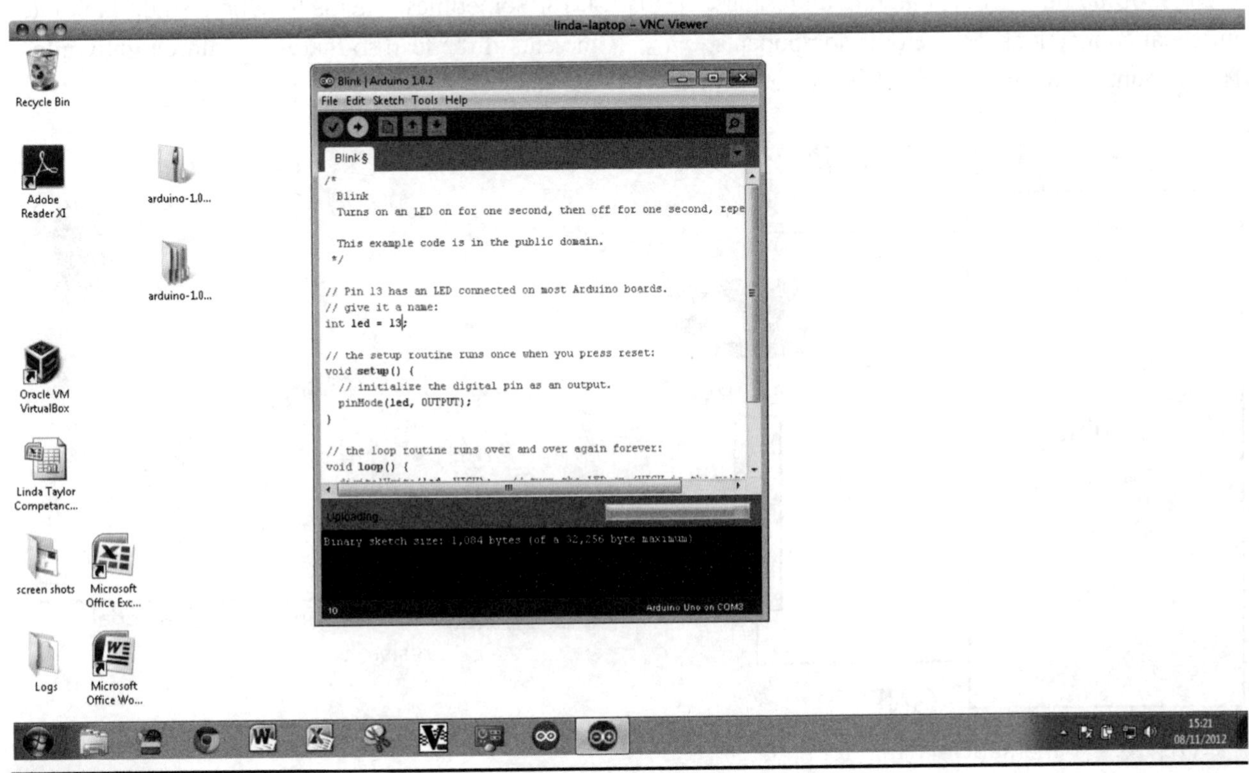

Figure 1-15 Uploading the sketch to the Arduino board.

LEDs cannot simply have voltage applied to them; they must have a current-limiting resistor attached. Both parts are readily available from any electronics suppliers. The component order codes for a number of suppliers are detailed in the Appendix.

The Arduino board connectors are designed to attach "shield" plug-in boards. However, for experimentation purposes, they also allow wires or component leads to be inserted directly into the sockets.

Figure 1-16 shows the schematic diagram for attaching the external LED.

This kind of schematic diagram uses special symbols to represent the electronic components. The LED appears rather like an arrow, which indicates that light-emitting diodes, in common with all diodes, only allow the current to flow in one direction. The little arrows next to the LED symbol indicate that it emits light.

The resistor is just depicted as a rectangle. Resistors are also often shown as a zigzag line. The rest of the lines on the diagram represent electrical connections between the components. These connections may be lengths of wire or tracks on a circuit board. In this case, they will just be the wires of the components.

We can connect the components directly to the Arduino sockets between the digital pin 12 and the GND pin, but first we need to connect one lead of the LED to one lead of the resistor.

It does not matter which lead of the resistor is connected to the LED; however, the LED must be connected the correct way. The LED will have one lead slightly longer than the other, and it is the longer lead that must be connected to digital pin 12 and the shorter lead that should be connected to the resistor. LEDs and some other components have the convention of making the positive lead longer than the negative one.

To connect the resistor to the short lead of the LED, gently spread the leads apart, and twist the short lead around one of the resistor leads, as shown in Figure 1-17.

Then push the LED's long lead into the digital pin 12 and the free lead of the resistor into one of the two GND sockets. This is shown in Figure 1-18. Sometimes it helps to bend a slight kink into the end of the lead so that it fits more tightly into the socket.

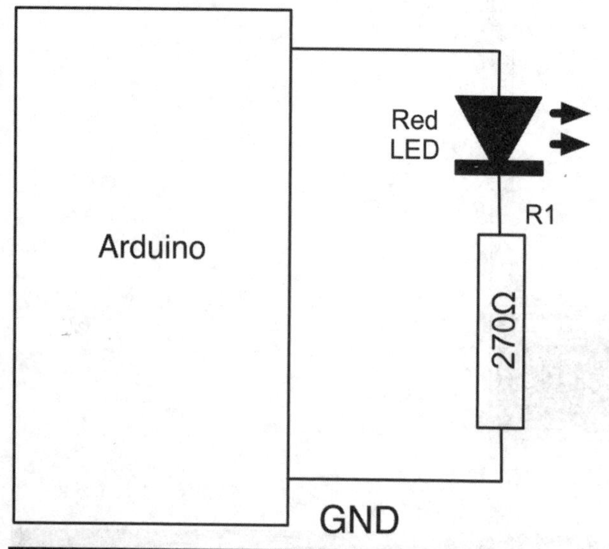

Figure 1-16 Schematic diagram for an LED connected to the Arduino board.

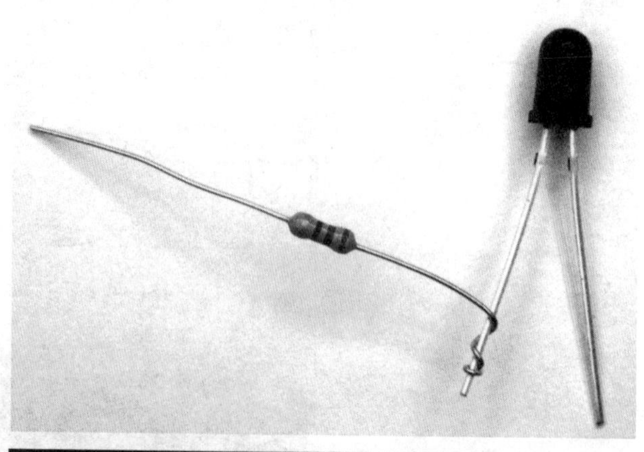

Figure 1-17 An LED connected to a serial resistor.

Figure 1-18 An LED connected to the Arduino board.

We can now modify our sketch to use the external LED that we have just connected. All we need to do is change the sketch so that it uses digital pin 12 instead of 13 for the LED. To do this, we change the line

```
int ledPin = 13;
 // LED connected to digital pin 13
```

to read

```
int ledPin = 12;
 // LED connected to digital pin 12
```

Now upload the sketch by clicking the Upload to IO Board button in the same way as you did when modifying the flash rate.

Breadboard

Twisting together a few wires is not practical for anything much more than a single LED. A breadboard allows us to build complicated circuits without the need for soldering. In fact, it is a good idea to build all circuits on a breadboard first to get the design right and then commit the design to solder once everything is working.

A breadboard comprises a plastic block with holes in it, with sprung metal connections behind. Electronic components are pushed through the holes at the front.

Underneath the breadboard holes, there are strips of connectors, so all the holes in a strip are connected together. The strips have a gap between them so that integrated circuits in dual-in-line packaging can be inserted without leads on the same row being shorted together.

We can build this project on a breadboard rather than with twisted wires. Figure 1-19 shows a photograph of this. Figure 1-20 makes it a little easier to see how the components are positioned and connected together.

You will notice that at the edges of the breadboard (top and bottom) there are two long horizontal strips. The connections on the back of these long strips run at right angles to the normal strips of connections and are used to provide power to the components on the breadboard. Normally, there is one for ground (0V or GND) and one for the positive supply voltage (usually 5V).

In addition to a breadboard, you will need some jumper wires (see the Appendix). These are short leads of a few inches in length of different colors. They are used to make connections between the

Figure 1-19 Project 1 on breadboard.

Arduino and the breadboard. Alternatively, you can use solid-core wire and some wire strippers or pliers to cut and remove the insulation from the ends of the wire. It is a good idea to have at least three different colors: red for all wires connected to the positive side of the supply, black for negative, and some other color (orange or yellow) for other connections. This makes it much easier to understand the layout of the circuit. You can also buy prepared short lengths of solid-core wire in a variety of colors. Note that it is not advisable to use multicore wire because it will tend to bunch up when you try to push it into the breadboard holes.

We can straighten out the wires of our LED and resistor and plug them into a breadboard. The breadboard used is often referred to as a half-size breadboard and has 30 rows of strips, each strip being five holes, then a gap, then another five

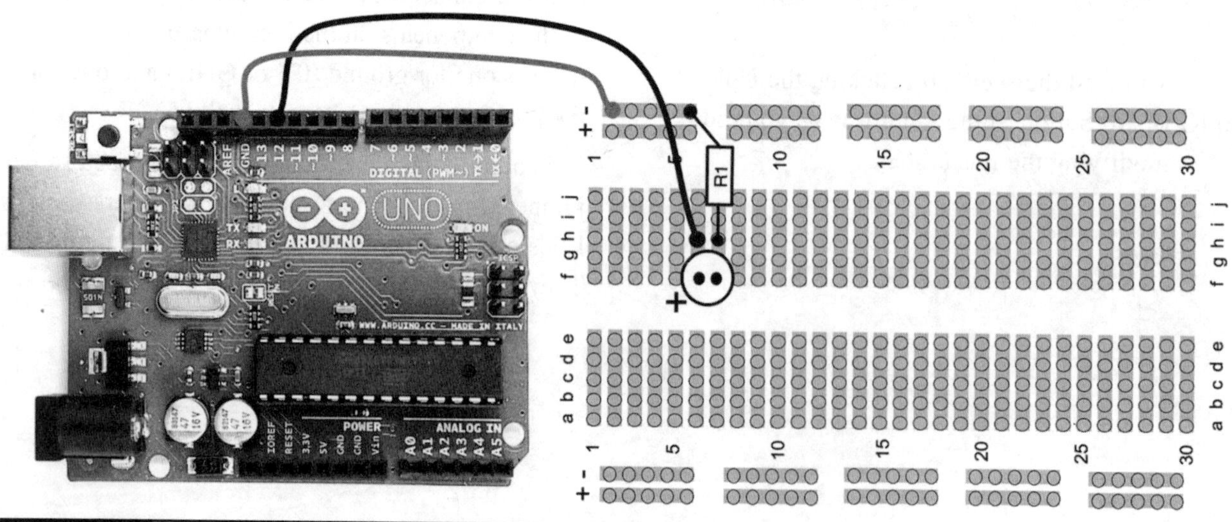

Figure 1-20 Project 1 breadboard layout.

holes. We will be using this breadboard a lot in this book, so if you can find something as similar as possible, it will make life easier. The actual board used was supplied by AdaFruit (see the Appendix), but it is a very common size and layout.

Summary

We have created our first project, albeit a very simple one. In Chapter 2 we will get a bit more background on the Arduino before moving on to some more interesting projects.

CHAPTER 2

A Tour of Arduino

In this chapter we look at the hardware of an Arduino board and also of the microcontroller at its heart. In fact, the board basically just provides support to the microcontroller, extending its pins to the connectors so that you can connect hardware to them and providing a USB link for downloading sketches, etc.

We also learn a few things about the C language used to program the Arduino, something we will build on in later chapters as we start on some practical project work.

Although this chapter gets quite theoretical at times, it will help you understand how your projects work. However, if you would prefer just to get on with your projects, you may wish to skim this chapter.

Microcontrollers

The heart of our Arduino is a microcontroller. Practically everything else on the board is concerned with providing the board with power and allowing it to communicate with your computer.

So what exactly do we get when we buy one of these little computers to use in our projects?

The answer is that we really do get a little computer on a chip. It has everything and more than the first home computers had. It has a processor, 2 or 2.5 kilobytes of random access memory (RAM) for holding data, 1 kilobyte of erasable programmable read-only memory (EPROM), and kilobytes of Flash memory for holding our programs. The important thing here is that this is kilobytes (thousands of bytes) not megabytes (millions of bytes) or gigabytes (billions of bytes). Most smart phones have upward of 1 gigabyte of memory. That is half a million times more RAM than an Arduino. An Arduino is a very feeble device indeed in terms of today's top-end hardware. However, this is not what the Arduino is for. The Arduino does not need to run a high-resolution screen or control complex networking. The Arduino is intended for much simpler control tasks.

Something that the Arduino has that you will not find on a smart phone is input and output pins. These input-output pins are what link the microcontroller to the rest of our electronics. This is what allows the Arduino to control things.

Inputs can read both digital (Is the switch on or off?) and analog (What is the voltage at a pin?). This enables us to connect many different types of sensors for light, temperature, sound, etc.

Outputs can also be analog or digital. So you can set a pin to be on or off (0V or 5V), and this can turn LEDs on and off directly, or you can use the output to control higher-power devices such as motors. They can also provide an analog output voltage. That is, you can set the output of a pin to some particular voltage, allowing you to control the speed of a motor or the brightness of a light, for example, rather than simply turning it on or off.

What's on an Arduino Board?

Figure 2-1 shows our Arduino board—in this case an Arduino Uno. Let us have a quick tour of the various components on the board.

Starting at the top next to the USB socket in the top left is the Reset button. Clicking this sends a logic pulse to the Reset pin of the microcontroller, causing the microcontroller to start its program afresh and clear its memory. Note that any program stored on the device will be retained because this is kept in nonvolatile Flash memory—that is, memory that remembers even when the device is not powered.

Power Supply

The Arduino can either be powered through the USB connection or the DC Barrel jack below it. When powering it from a DC adaptor or batteries, anything between 7.5V and 12V DC can be supplied through the power socket.

When the Arduino is powered, the power LED on the right of the Uno (left of the Leonard) will be lit.

Power Connections

Next, let us look at the connectors at the bottom of Figure 2-1. Apart from the first connection, you can read the connection names next to the connectors.

The first unlabeled connection is reserved for later use. The next pin, "IOREF," is used to indicate the voltage at which the Arduino operates. Both the Uno and Leonardo operate at 5V, so this pin will always be at 5V, and we will not be using it for anything. Its purpose is to allow shields attached to 3V Arduinos such as the Arduino Due to detect the voltage at which the Arduino operates.

The next connection is Reset. This does the same thing as pressing the Reset button on the Arduino. Rather like rebooting a PC, it resets the

Figure 2-1 Components of an Arduino Uno.

microcontroller, beginning its program from the start. The Reset connector allows you to reset the microcontroller by momentarily setting this pin high (connecting it to +5V).

The rest of the pins in this section provide different voltages (3.3V, 5V, GND, and 9V), as labeled. GND, or ground, just means 0V. It is the reference voltage to which all other voltages on the board are relative.

At this point it would be useful to remind the reader about the difference between voltage and current. There is no perfect analogy for the behavior of electrons in a wire, but the author finds an analogy with water in pipes to be helpful, particularly in dealing with voltage, current, and resistance. The relationship between these three things is called *Ohm's law*.

Figure 2-2 summarizes the relationship between voltage, current, and resistance. The left side of the diagram shows a circuit of pipes, where the top of the diagram is higher up (in elevation) than the bottom of the diagram. So water will naturally flow from the top of the diagram to the bottom. Two factors determine how much water passes any point in the circuit in a given time (the current):

- The height of the water (or, if you prefer, the pressure generated by the pump). This is like voltage in electronics.
- The resistance to flow offered by the constriction in the pipe work.

The more powerful the pump, the higher the water can be pumped and the greater the current that will flow through the system. On the other hand, the greater the resistance offered by the pipe work, the lower the current.

In the right half of Figure 2-2, we can see the electronic equivalent of our pipe work. In this case, current is actually a measure of how many electrons flow past a point per second. And yes, resistance is the resistance to the flow of electrons.

Instead of height or pressure, we have a concept of voltage. The bottom of the diagram is at 0V, or ground, and we have shown the top of the diagram as being at 5V. So the current that flows (I) will be the voltage difference (5) divided by the resistance (R).

Ohm's law is usually written as $V = IR$. Normally, we know what V is and are trying to calculate R or I, so we can do a bit of rearranging to have the more convenient $I = V/R$ and $R = V/I$.

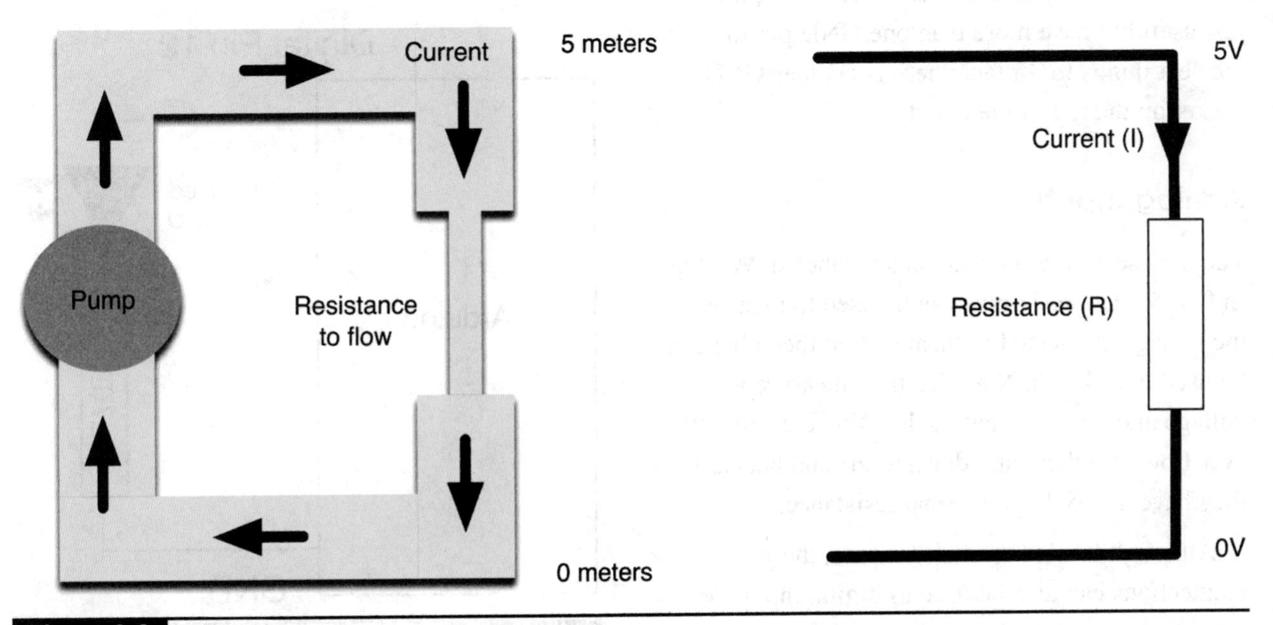

Figure 2-2 Ohm's law.

It is very important to do a few calculations using Ohm's law when connecting things to your Arduino, or you may damage it if you ask it to supply too much current. Generally, though, the Arduino boards are remarkably tolerant of accidental abuse.

So going back to our Arduino power pins, we can see that the Arduino board will supply us with useful voltages of 3.3V and 5V. If the Arduino is supplied with a higher voltage through the power jack, then this voltage will also be available on the Vin pin. We can use any of those supplies to cause a current to flow, as long as we are careful not to make it a short circuit (no resistance to flow), which would cause a potentially large current to flow that could cause damage. In other words, we have to make sure that anything we connect to the supply has enough resistance to prevent too much current from flowing. As well as supplying a particular voltage, each of those supply connections will have a maximum current that can be allowed to flow. Those currents are 50 mA (thousandths of an ampere) for the 3.3V supply, and although it is not stated in the Arduino specification, probably around 300 mA for the 5V supply.

The two GND connections are identical; it is just useful to have more than one GND pin to connect things to. In fact, there is another GND socket on the top of the board.

Analog Inputs

The next section of connections is labeled "Analog In 0 to 5." These six pins can be used to measure the voltage connected to them so that the value can be used in a sketch. Note that they measure a voltage and not a current. Only a tiny current will ever flow into them and down to ground because they have a very large internal resistance.

Although labeled as analog inputs, these connections can also be used as digital inputs or outputs, but by default, they are analog inputs.

Unlike the Uno, the Leonardo can also use digital pins 4, 6, 8, 9, 10, and 12 as analog inputs.

Digital Connections

We now switch to the top connector and start on the right side (Figure 2-1). We have pins labeled "Digital 0 to 13." These can be used as either inputs or outputs. When using them as outputs, they behave rather like the supply voltages we talked about earlier, except that these are all 5V and can be turned on or off from our sketch. So, if we turn them on from our sketch, they will be at 5V, and if we turn them off, they will be at 0V. As with the supply connectors, we have to be careful not to exceed their maximum current capabilities.

These connections can supply 40 mA at 5V. That is more than enough to light a standard LED but not enough to drive an electric motor directly.

As an example, let us look at how we would connect an LED to one of these digital connections. In fact, let's go back to Project 1 in Chapter 1.

As a reminder, Figure 2-3 shows the schematic diagram for driving the LED that you first used in Chapter 1. If you were to not use a resistor with

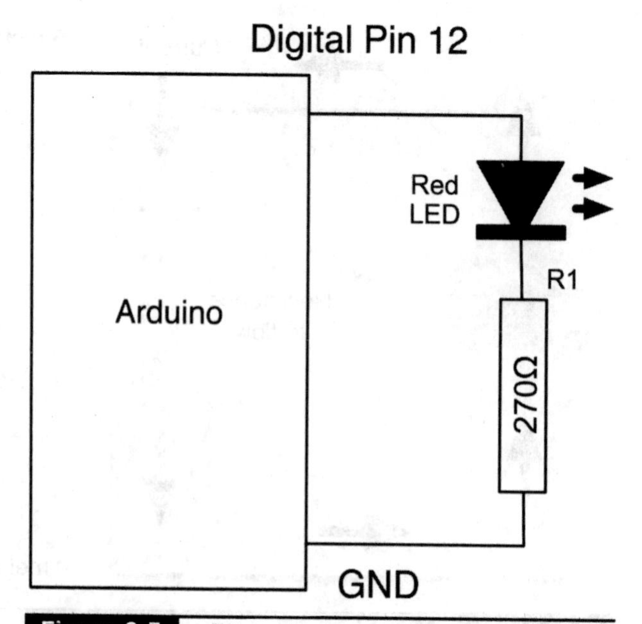

Figure 2-3 LED and series resistor.

our LED but simply connect the LED between pin 12 and GND, then when you turned digital output 12 on (5V), you would burn out the LED, destroying it.

This is so because LEDs have a very low resistance and will cause a very high current to flow unless they are protected from themselves by using a resistor to limit the flow of current.

An LED needs about 10 mA to shine reasonably brightly. The Arduino can supply 40 mA, so there is no problem there; we just need to choose a sensible value of resistor.

LEDs have the interesting property that no matter how much current flows through them, there will always be about 2V between their pins. We can use this fact and Ohm's law to work out the right value of resistor to use.

We know that (at least when it's on) the output pin will be supplying 5V. Now we have just said that 2V will be "dropped" by our LED, leaving 3V (5 – 2) across our current-limiting resistor. We want the current flowing around the circuit to be 10 mA, so we can see that the value for the resistor should be

$R = V/I$

$R = 3V/10\ mA$

$R = 3V/0.01\ A$

$R = 300\ \Omega$

Resistors come in standard values, and the closest value to 300 Ω is 270 Ω. This means that instead of 10 mA, the current will actually be

$I = V/R$

$I = 3/270$

$I = 11.111\ mA$

These things are not critical, and the LED probably would be equally happy with anything between 5 and 30 mA, so 270 Ω will work just fine as would 220 Ω or 330 Ω.

We can also set one of these digital connections to be an input, in which case it works rather like an analog input, except that it will just tell us if the voltage at a pin is above a certain threshold (roughly 2.5V) or not.

Some of the digital connections (3, 5, 6, 9, 10, and 11) have the letters "PWM" next to them. These can be used to provide a variable output voltage rather than a simple 5V or nothing.

On the left side of the top connector in Figure 2-1, there is another GND connection and a connection called "AREF." AREF can be used to scale the readings for analog inputs. This is not used in this book.

Digital pin 13 is also connected to an LED known as the "L" LED.

Microcontroller

Getting back to our tour of the Arduino board, the microcontroller chip itself is the black rectangular device with 28 pins. This is fitted into a dual in-line (DIL) socket so that it can be easily replaced. The 28-pin microcontroller chip used on an Arduino Uno is the ATmega328. The biggest difference between the Uno and the Leonardo (Figure 2-4) is that the Leonardo has a surface-mount chip permanently soldered into place. This effectively makes it very hard to replace the microcontroller if it becomes damaged.

The Leonardo also uses a different version of the microcontroller board that includes the USB interface circuitry that is separate in the Uno.

This makes the Leonardo board more sparsely populated with components and is a reason for its lower cost. Figure 2-5 is a block diagram showing the main features of the ATmega328 microcontroller chip.

Figure 2-4 The Arduino Leonardo.

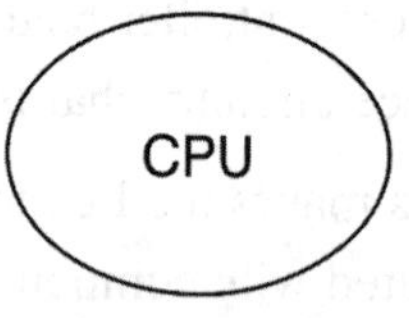

Figure 2-5 ATmega328 block diagram.

The heart, or perhaps more appropriately, the brain, of the device is the central processing unit (CPU). It controls everything that goes on within the device. It fetches program instructions stored in the Flash memory and executes them. This might involve fetching data from working memory (RAM), changing it, and then putting it back. Or it may mean changing one of the digital outputs from 0V to 5V.

The electrically erasable programmable read-only memory (EEPROM) memory is a little like the Flash memory in that it is nonvolatile. That is, you can turn the device off and on, and it will not have forgotten what is in the EEPROM. Whereas the Flash memory is intended for storing program instructions (from sketches), the EEPROM is used to store data that you do not want to lose in the event of a reset or power failure.

The Leonardo's microcontroller is similar, except that it has 2.5 kilobytes of RAM rather than 2 kilobytes.

Other Components

Above and to the left of the microcontroller there is a small silver rectangular component. This is a quartz crystal oscillator. It "ticks" 16 million times a second, and on each of those ticks, the microcontroller can perform one operation—an addition, subtraction, etc.

To the right of the microcontroller chip is the serial programming connector (ICSP header). It offers another means of programming the Arduino without using the USB port. Since we do have a USB connection and software that makes it convenient to use, we will not avail ourselves of this feature.

In the top left of the board next to the USB socket is the USB interface chip. This converts the signal levels used by the USB standard to levels that can be used directly by the Arduino board.

The Arduino Family

It's useful to have a little background on the Arduino boards. We will be using the Uno or Leonardo for most of our projects; however, we will also dabble with the interesting Lilypad Arduino.

The Lilypad (Figure 2-6) is a tiny, thin Arduino board that can be stitched into clothing for applications that have become known as wearable computing. It does not have a USB connection, and you must use a separate adaptor to program it. This is an exceptionally beautiful design. Inspired by its clocklike appearance, we will use this in Project 29 (Unfathomable Binary Clock).

At the other end of the spectrum is the Arduino Mega. This board has a faster processor with more memory and a greater number of input-output pins.

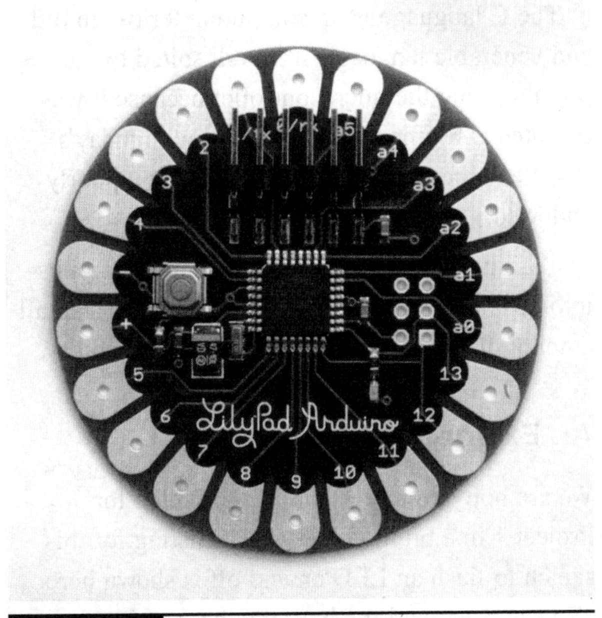

Figure 2-6 Arduino Lilypad.

Cleverly, the Arduino Mega can still use shields built for the smaller Arduino Uno and Leonardo boards, which sit at the front of the board, allowing access to the double row of connectors for the Mega's additional connections at the rear. Only the most demanding of projects really need an Arduino Mega.

Taking this to the next stage is the Arduino Due. This Arduino board is the same size as an Arduino Mega but has a much more powerful processor, 96 kilobytes of RAM, and 512 megabytes of Flash memory and is clocked at 84 MHz rather than the Uno's 16 MHz.

The C Language

Many languages are used to program microcontrollers, from hard-core Assembly language to graphical programming languages such as Flowcode. Arduino sits somewhere in between these two extremes and uses the C programming language. It does, however, wrap up the C language, hiding away some of the complexity. This makes it easy to get started.

The C language is, in computing terms, an old and venerable language. It is well suited to programming the microcontroller because it was invented at a time when compared with today's monsters, the typical computer was quite poorly endowed.

C is an easy language to learn, yet it compiles into efficient machine code that only takes a small amount of room in our limited Arduino memory.

An Example

We are now going to examine the sketch for Project 1 in a bit more detail. The listing for this sketch to flash an LED on and off is shown here. You can ignore all the lines that begin with // or blocks of lines that start with /* and end with */ because these are comment lines that have no effect on the program and are just there for information.

```
int ledPin = 13;
  // LED connected to digital pin 13
void setup()
{
  pinMode(ledPin, OUTPUT);
}

void loop()
{
  digitalWrite(ledPin, HIGH);
    // set the LED on
  delay(1000);
    // wait for a second
  digitalWrite(ledPin, LOW);
    // set the LED off
  delay(1000);
    // wait for a second
}
```

Also, it is a good idea to include comments that describe a tricky bit of code or anything that requires some explanation.

The Arduino development environment uses something called a *compiler* that converts the script into the machine code that will run on the microcontroller.

So, moving onto the first real line of code, we have

```
int ledPin = 13;
```

This line of code gives a name to the digital output pin that we are going to connect to the LED. If you look carefully at your Arduino board, you will see the connector for pin 13 between GND and pin 12 on the Arduino's top connector. The Arduino board has a small LED already soldered onto the board and connected to pin 13. We are going to change the voltage of this pin to between 0V and 5V to make the LED flash.

We are going to use a name for the pin so that it's easy to change it and use a different one. You can see that we refer to ledPin later in the sketch. You may prefer to use pin 12 and the external LED that you used with your breadboard in Chapter 1. But for now we will assume that you are using the built-in LED attached to pin 13.

You will notice that we did not just write

```
led pin = 13
```

This is so because compilers are kind of fussy and precise about how we write our programs. Any name we use in a program cannot use spaces, so it is a convention to use what is called *bumpy case*. So we start each word (apart from the first) with an uppercase letter and remove the space; that gives us

```
ledPin = 13
```

The word ledPin is what is termed a *variable*. When you want to use a variable for the first time in a sketch, you have to tell the compiler what type of variable it is. It may be an int, as is the case here, or a float, or a number of other types that we will describe later in this chapter.

An int is an *integer*—that is, a whole number—which is just what we need when referring to a particular pin on the Arduino. There is, after all, no pin 12.5, so it would not be appropriate to use a floating-point number (float).

The syntax for a variable declaration is

```
type variableName = value;
```

So first we have the type (int), then a space, then a variable name in bumpy case (ledPin), then an equal sign, then a value, and finally, a semicolon to indicate the end of the line:

```
int ledPin = 13;
```

As I mentioned, the compiler is fussy, so if you forget the semicolon, you will receive an error message when you compile the sketch. Try removing the semicolon and clicking the Play button. You should see a message like this:

```
error: expected unqualified-id before
numeric constant
```

It's not exactly "you forgot a semicolon," and it is not uncommon for error messages to be similarly misleading.

The compiler is much less fussy about "whitespace" characters, that is, spaces, tabs, and the return character. So if you omitted the spaces either side of the = sign, it would still compile. Use of spaces and tabs makes the code easier to read, and by sticking to a convention and always formatting your code the same and in a fairly standard way, you will make it much easier for other people to understand your code.

The next lines of the sketch are

```
void setup()
  // run once, when the sketch starts
{
  pinMode(ledPin, OUTPUT);
  // sets the digital pin as output
}
```

This is what is called a *function*, and in this case, the function is called setup. Every sketch must contain a setup function, and the lines of code inside the function surrounded by curly brackets will be carried out in the order that they are written. In this case, that is just the line starting with pinMode.

A good starting point for any new project is to copy this example project and then alter it to your needs.

We will not worry too much about functions at this stage, other than to say that the setup function will be run every time the Arduino is reset, including when the power is first turned on. It will also be run every time a new sketch is uploaded.

In this case, the only line of code in setup is

```
pinMode(ledPin, OUTPUT);
  // sets the digital pin as output
```

The line can be thought of as a command to the Arduino to use the ledPin as a digital output. If we had a switch connected to ledPin, we could set it as an input using

```
pinMode(ledPin, INPUT);
```

However, we would call the variable something more appropriate, such as switchPin.

The words INPUT and OUTPUT are what are called *constants*. They will actually be defined within C to be a number. INPUT may be defined as 0 and OUPUT as 1, but you never need to actually see what number is used because you always refer to them as INPUT or OUTPUT. Later in this chapter we will see two more constants, HIGH and LOW, that are used when setting the output of a digital pin to +5V or 0V, respectively.

The next section of code is another function that every Arduino sketch must have; it is called loop:

```
void loop()
{
  digitalWrite(ledPin, HIGH);
  // sets the LED on
  delay(1000);
  // waits for a second
  digitalWrite(ledPin, LOW);
  // sets the LED off
  delay(1000);
  // waits for a second
}
```

The function loop will be run continuously until the Arduino is powered down. That is, as soon as it finishes executing the commands it contains, it will begin again. Remember that an Arduino board is capable of running 16 million commands per second, so things inside the loop will happen frequently if you let them.

In this case, what we want the Arduino to keep doing continuously is to turn the LED on, wait a second, turn the LED off, and then wait another second. When it has finished doing this, it will begin again, turning the LED on. In this way it will go round the loop forever.

By now, the command syntax for digitalWrite and delay will be becoming more familiar. Although we can think of them as commands that are sent to the Arduino board, they are actually functions just like setup and loop, but in this case they have what are called *parameters*. The parameters are enclosed by parentheses and separated by commas. In the case of digitalWrite, it is said to take two parameters: the Arduino pin to write to and the value to write.

In our example, we pass the parameters of ledPin and HIGH to turn the LED on and then ledPin and LOW to turn it off again.

Variables and Data Types

We have already met the variable ledPin and declared it to be of type int. Most of the variables that you use in your sketches are also likely to be of type int. An int holds an integer number between –32,768 and +32,767. This uses just 2 bytes of data for each number stored from the 1024 available bytes of storage on an Arduino. If that range is not enough, you can use a long, which uses 4 bytes for each number and will give you a range of numbers from –2,147,483,648 to +2,147,483,647.

Most of the time, an int represents a good compromise between precision and use of memory.

If you are new to programming, I would use int for almost everything and gradually expand your repertoire of data types as your experience grows.

Other data types available to you are summarized in Table 2-1.

One thing to consider is that if data types exceed their range, strange things happen. So, if you have a byte variable with 255 in it and you add 1 to it, you get 0. More alarmingly, if you have an int variable with 32,767 and you add 1 to it, you will end up with –32,768.

Until you are completely happy with these different data types, I would recommend sticking to it because it works for practically everything.

Arithmetic

It is fairly uncommon to need to do much in the way of arithmetic in a sketch. Occasionally, you will need to do a bit of scaling of, say, an analog input to turn it into a temperature or, more typically, add 1 to a counter variable.

When you are performing some calculation, you need to be able to assign the result of the calculation to a variable.

TABLE 2-1 Data Types in C

Type	Memory (bytes)	Range	Notes
boolean	1	True or false (0 or 1)	
char	1	-128 to +128	Used to represent an ASCII character code (e.g., A is represented as 65). Its negative numbers are not normally used.
byte	1	0 to 255	
int	2	-32,768 to +32,767	
unsigned int	2	0 to 65,536	Can be used for extra precision where negative numbers are not needed. Use with caution because arithmetic with int may cause unexpected results.
long	4	-2,147,483,648 to 2,147,483,647	Needed only for representing very large numbers.
unsigned long	4	0 to 4,294,967,295	See unsigned int.
float	4	-3.4028235E+38 to +3.4028235E+38	
double	4	as float	Normally, this would be 8 bytes and higher precision than float with a greater range. However, on Arduino, it is the same as float.

The following lines of code contain two assignments. The first gives the variable y the value 50, and the second gives the variable x the value of y + 100.

```
y = 50;
x = y + 100;
```

Strings

When programmers talk of *strings*, they are referring to a string of characters such as the much-used message “Hello World.” In the world of Arduino, there are a couple of situations where you might want to use strings: when writing messages to an LCD display or sending back serial text data over the USB connection.

Strings are created using the following syntax:

```
char* message = "Hello World";
```

The char* word indicates that the variable message is a pointer to a character. For now, we do not need to worry too much about how this works. We will meet this later in the book when we look at interfacing with textual LCD displays.

Conditional Statements

Conditional statements are a means of making decisions in a sketch. For instance, your sketch may turn the LED on if the value of a temperature variable falls below a certain threshold.

The code for this is shown here:

```
if (temperature < 15)
{
  digitalWrite(ledPort, HIGH);
}
```

The line or lines of code inside the curly braces will only be executed if the condition after the if keyword is true.

The condition has to be contained in parentheses and is what programmers call a *logical expression*. A logical expression is like a mathematical sentence that must always return one of two possible values: true or false.

The following expression will return true if the value in the temperature variable is less than 15:

```
(temperature < 15)
```

As well as <, you have: >, <=, and >=. To see if two numbers are equal, you can use ==, and to test if they are not equal, you can use !=.

So the following expression would return true if the temperature variable had a value that was anything except 15:

```
(temperature != 15)
```

You can also make complex conditions using what are called *logical operators*. The principal operators being && (and) and || (or).

So an example that turned the LED on if the temperature was less than 15 or greater than 20 might look like this:

```
if ((temperature < 15) || (temperature
    > 20))
{
  digitalWrite(ledPort, HIGH);
}
```

Often, when using an if statement, you want to do one thing if the condition is true and a different thing if it is false. You can do this by using the else keyword, as shown in the following example. Note the use of nested parentheses to make it clear what is being or'd with what.

```
if ((temperature < 15) ||
      (temperature > 20))
{
  digitalWrite(ledPort, HIGH);
}
else
{
  digitalWrite(ledPort, LOW);
}
```

Summary

In this chapter we have explored the hardware provided by the Arduino and refreshed our knowledge of a little elementary electronics.

We have also started our exploration of the C programming language. Don't worry if you found some of this hard to follow. There is a lot to take in if you are not familiar with electronics, and while the author's goal is to explain how everything works, you are completely at liberty to simply start on the projects first and come back to the theory when you are ready.

If you want to learn more about programming the Arduino in C, then the book *Programming Arduino: Getting Started with Sketches* by this author is devoted to that topic.

In Chapter 3 we will come to grips with programming our Arduino board and embark on some more serious projects.

CHAPTER 3

LED Projects

IN THIS CHAPTER WE ARE GOING TO start building some LED-based projects. We will keep the hardware fairly simple so that we can concentrate on the programming of the Arduino.

Programming microcontrollers can be a tricky business requiring an intimate knowledge of the inner workings of the device: fuses, registers, etc. This is due, in part, to the fact that modern microcontrollers are almost infinitely configurable. Arduino standardizes its hardware configuration, which, in return for a small loss of flexibility, makes the devices a great deal easier to program.

Project 2
Morse Code S.O.S. Flasher

Morse code used to be a vital method of communication in the 19th and 20th centuries. Its coding of letters as a series of long and short dots meant that it could be sent over telegraph wires, over a radio link, and using signaling lights. The letters S.O.S. (Save Our Souls) are still recognized as an international signal of distress.

In this project, we will make our LED flash the sequence S.O.S. over and over again.

For this project you will need just the same components as for Project 1.

COMPONENTS AND EQUIPMENT

	Description	Appendix
	Arduino Uno or Leonardo	m1/m2
D1	5-mm red LED	s1
R1	270 Ω, 0.25 W resistor	r3

- Almost any commonly available LED and 270 Ω resistor will be fine.
- No tools other than a pair of pliers or wire cutters are required.

Hardware

The hardware is exactly the same as for Project 1. So you can either just plug the resistor and LED directly into the Arduino connectors or use a breadboard (see Chapter 1).

Software

Rather than start typing this project in from scratch, we will use Project 1 as a starting point. So please complete Project 1 before you begin this project.

If you have not already done so, download the project code from www.arduinoevilgenius.com; then you can also just load the completed sketch for Project 1 from your Arduino Sketchbook and download it to the board (see Chapter 1). However, it will help you to understand Arduino better if you modify the sketch from Project 1 as suggested next.

Modify the loop function of Project 1 so that it now appears as shown here. Note that copy and paste are highly recommended in this kind of situation.

```
void loop()
{
  digitalWrite(ledPin, HIGH);
  // S (...) first dot
  delay(200);
  digitalWrite(ledPin, LOW);
  delay(200);
  digitalWrite(ledPin, HIGH);
  // second dot
  delay(200);
  digitalWrite(ledPin, LOW);
  delay(200);
  digitalWrite(ledPin, HIGH);
  // third dot
  delay(200);
  digitalWrite(ledPin, LOW);
  delay(500);
  digitalWrite(ledPin, HIGH);
  // O (—) first dash
  delay(500);
  digitalWrite(ledPin, LOW);
  delay(500);
  digitalWrite(ledPin, HIGH);
  // second dash
  delay(500);
  digitalWrite(ledPin, LOW);
  delay(500);
  digitalWrite(ledPin, HIGH);
  // third dash
  delay(500);
  digitalWrite(ledPin, LOW);
  delay(500);
  digitalWrite(ledPin, HIGH);
  // S (...) first dot
  delay(200);
  digitalWrite(ledPin, LOW);
  delay(200);
  digitalWrite(ledPin, HIGH);
  // second dot
  delay(200);
  digitalWrite(ledPin, LOW);
  delay(200);
  digitalWrite(ledPin, HIGH);
  // third dot
  delay(200);
  digitalWrite(ledPin, LOW);
  delay(1000);
  // wait 1 second before we start
     again
}
```

This would all work, and feel free to try it; however, we are not going to leave it there. We are going to alter our sketch to improve it and at the same time make it a lot shorter.

We can reduce the size of the sketch by creating our own function to replace the four lines of code involved in any flash of the LED with one line.

After the loop function's final curly brace, add the following code:

```
void flash(int duration)
{
  digitalWrite(ledPin, HIGH);
  delay(duration);
  digitalWrite(ledPin, LOW);
  delay(duration);
}
```

Now modify the loop function so that it looks like this:

```
void loop()
{
  flash(200); flash(200); flash(200);
  // S
  delay(300);
  // otherwise the flashes run together
  flash(500); flash(500); flash(500);
  // O
  flash(200); flash(200); flash(200);
  // S
  delay(1000);
  // wait 1 second before we start
     again
}
```

The whole final listing is shown in Listing Project 2.

LISTING PROJECT 2

```
int ledPin = 13;

void setup()                          // run once, when the sketch starts
{
  pinMode(ledPin, OUTPUT);            // sets the digital pin as output
}

void loop()
{
  flash(200); flash(200); flash(200);   // S
  delay(300);                           // otherwise the flashes run together
  flash(500); flash(500); flash(500);   // O
  flash(200); flash(200); flash(200);   // S
  delay(1000);                          // wait 1 second before we start again
}

void flash(int duration)
{
  digitalWrite(ledPin, HIGH);
  delay(duration);
  digitalWrite(ledPin, LOW);
  delay(duration);
}
```

This makes the sketch a lot smaller and a lot easier to read.

Putting It All Together

That concludes Project 2. We will now cover some more background on programming the Arduino before we go on to look at Project 3, where we will use our same hardware to write a Morse code translator, with which we can type sentences on our computer and have them flashed as Morse code. In Project 4 we will improve the brightness of our flashing by replacing our red LED with a high-power Luxeon-type LED.

But first we need a little more theory in order to understand Projects 3 and 4.

Loops

Loops allow us to repeat a group of commands a certain number of times or until some condition is met. In Project 2, we only want to flash three dots for an S, so it is no great hardship to repeat the flash command three times. However, it would be far less convenient if we needed to flash the LED 100 or 1000 times. In that case we can use the for language command in C:

```
for (int i = 0; i < 100; i ++)
{
  flash(200);
}
```

The for loop is a bit like a function that takes three arguments, although here those arguments are separated by semicolons rather than by the usual commas. This is just a quirk of the C language. The compiler will soon tell you when you get it wrong.

The first thing in the parentheses after for is a variable declaration. This specifies a variable to be used as a counter variable and gives it an initial value—in this case 0.

The second part is a condition that must be true for us to stay in the loop. In this case we will stay in the loop as long as i is less than 100, but as soon as i is 100 or more, we will stop doing the things inside the loop.

The final part is what to do every time you have done all the things in the loop. In this case, that is increment i by 1 so that it can, after 100 trips around the loop, cease to be less than 100 and cause the loop to exit.

Another way of looping in C is to use the while command. The same example shown previously could be accomplished using a while command, as shown here:

```
int i = 0;
while (i < 100)
{
  flash(200);
  i ++;
}
```

The expression in parentheses after the while must be true to stay in the loop. When it is no longer true, the sketch will continue running the commands after the final curly brace.

The curly braces are used to bracket together a group of commands. In programming parlance, they are known as a *block*.

Arrays

Arrays are a way of containing a list of values. The variables we have met so far have only contained a single value, usually an int. By contrast, an array contains a list of values, and you can access any one of those values by its position in the list.

C, like the majority of programming languages, begins its index positions at 0 rather than 1. This means that the first element is actually element zero.

To illustrate the use of arrays, we could change our Morse code example to use an array of flash durations. We can then use a for loop to step through each of the items in the array.

First let's create an array of type int containing the durations:

```
int durations[] = {200, 200, 200, 500,
  500, 500, 200, 200, 200}
```

You indicate that a variable contains an array by placing [] after the variable name. If you are setting the contents of the array at the same time you are defining it, as in the preceding example, you do not need to specify the size of the array. If you are not setting its initial contents, then you need to specify the size of the array inside the square brackets. For example:

```
int durations[10];
```

Now we can modify our loop method to use the array:

```
void loop()
  // run over and over again
{
  for (int i = 0; i < 9; i++)
  {
    flash(durations[i]);
  }
  delay(1000);
  // wait 1 second before we start
  // again
}
```

An obvious advantage of this approach is that it is easy to change the message by simply altering the durations array. In Project 3 we will take the use of arrays a stage further to make a more general-purpose Morse code flasher.

Project 3
Morse Code Translator

In this project we are going to use the same hardware as for Projects 1 and 2, but we are going to write a new sketch that will let us type in a sentence on our computer and have our Arduino board convert that into the appropriate Morse code dots and dashes.

Figure 3-1 shows the Morse code translator in action. The contents of the message box are being flashed as dots and dashes on the LED.

To do this, we will make use of what we have learned about arrays and strings and also learn something about sending messages from our computer to the Arduino board through the USB cable.

For this project, you will need just the same components as for Projects 1 and 2. In fact, the hardware is exactly the same; we are just going to modify the sketch of Project 1.

COMPONENTS AND EQUIPMENT

	Description	Appendix
	Arduino Uno or Leonardo	m1/m2
D1	5-mm red LED	s1
R1	270 Ω, 0.25 W resistor	r3

Figure 3-1 Morse code translator.

Hardware

Please refer back to Project 1 for the hardware construction for this project.

You can either just plug the resistor and LED directly into the Arduino connectors or use the breadboard (see Chapter 1). You can even just change the ledPin variable in the sketch to be pin 13 so that you use the built-in LED and do not need any external components at all.

Software

The letters in Morse code are shown in Table 3-1.

TABLE 3-1 Morse Code Letters

A	.-	N	-.	0	-----
B	-...	O	---	1	.----
C	-.-.	P	.--.	2	..---
D	-..	Q	--.-	3	...--
E	.	R	.-.	4	....-
F	..-.	S	...	5	.....
G	--.	T	-	6	-....
H	....	U	..-	7	--...
I	..	V	...-	8	---..
J	.---	W	.--	9	----.
K	-.-	X	-..-		
L	.-..	Y	-.--		
M	--	Z	--..		

Some of the rules of Morse code are that a dash is three times as long as a dot, the time between each dash or dot is equal to the duration of a dot, the space between two letters is the same length as a dash, and the space between two words is the same duration as seven dots.

For the sake of this project, we will not worry about punctuation, although it would be an interesting exercise for you to try adding this to the sketch. For a full list of all the Morse characters, see http://en.wikipedia.org/wiki/Morse_code.

The sketch for this is shown in Listing Project 3. An explanation of how it all works follows.

LISTING PROJECT 3

```
int ledPin = 12;

char* letters[] = {
  ".-", "-...", "-.-.", "-..", ".", "..-.", "--.", "....", "..",       // A-I
  ".---", "-.-", ".-..", "--", "-.", "---", ".--.", "--.-", ".-.",    // J-R
  "...", "-", "..-", "...-", ".--", "-..-", "-.--", "--.."           // S-Z
};

char* numbers[] = {"-----", ".----", "..---", "...--", "....-", ".....", "-....",
  "--...", "---..", "----."};

int dotDelay = 200;

void setup()
{
  pinMode(ledPin, OUTPUT);
  Serial.begin(9600);
}

void loop()
```

LISTING PROJECT 3 ***(continued)***

```
{
  char ch;
  if (Serial.available())           // is there anything to be read from USB?
  {
    ch = Serial.read();             // read a single letter
    if (ch >= 'a' && ch <= 'z')
    {
       flashSequence(letters[ch - 'a']);
    }
    else if (ch >= 'A' && ch <= 'Z')
    {
      flashSequence(letters[ch - 'A']);
    }
    else if (ch >= '0' && ch <= '9')
    {
      flashSequence(numbers[ch - '0']);
    }
    else if (ch == ' ')
    {
    delay(dotDelay * 4);            // gap between words
    }
  }
}

void flashSequence(char* sequence)
{
  int i = 0;
  while (sequence[i] != NULL)
  {
      flashDotOrDash(sequence[i]);
      i++;
  }
  delay(dotDelay * 3);              // gap between letters
}

void flashDotOrDash(char dotOrDash)
{
  digitalWrite(ledPin, HIGH);
  if (dotOrDash == '.')
  {
    delay(dotDelay);
  }
  else // must be a -
  {
    delay(dotDelay * 3);
  }
  digitalWrite(ledPin, LOW);
  delay(dotDelay);                  // gap between flashes
}
```

We keep track of our dots and dashes using arrays of strings. We have two of these, one for letters and one for numerals. So, to find out what we need to flash for the first letter of the alphabet (A), we will get the string letters[0]—remember, the first element of an array is element 0, not element 1.

The variable dotDelay is defined, so if we want to make our Morse code flash faster or slower, we can change this value because all the durations are defined as multiples of the time for a dot.

The setup function is much the same as for our earlier projects; however, this time we are getting communications from the USB port, so we must add the command

```
Serial.begin(9600);
```

This tells the Arduino board to set the communications speed through the USB to be 9600 baud. This is not very fast, but fast enough for our Morse code messages. It is also a good speed to set it to because that is the default speed used by the Arduino software on your computer.

In the loop function we are going to repeatedly see if we have been sent any letters over the USB connection and if we have to process the letter. The Arduino function Serial.available() will be true if there is a character to be turned into Morse code, and the Serial.read() function will give us that character, which we assign to a variable called ch that we defined just inside the loop.

We then have a series of *if* statements that determine whether the character is an uppercase letter, a lowercase letter, or a space character separating two words. Looking at the first *if* statement, we are testing to see if the character's value is greater than or equal to a and less than or equal to z. If that is the case, we can find the sequence of dashes and dots to flash using the letters array that we defined at the top of the sketch. We determine which sequence from the array to use by subtracting a from the character in ch.

At first sight, it might look strange to be subtracting one letter from another, but it is perfectly acceptable to do this in C. So, for example, a - a is 0, whereas d - a will give us the answer 3. So, if the letter that we read from the USB connections were f, we would calculate f - a, which gives us 5 as the position of the letters array. Looking up letters[5] will give us the string ..-.. and we pass this string to a function called flashSequence.

The flashSequence function is going to loop over each of the parts of the sequence and flash it as either a dash or a dot. Strings in C all have a special code on the end of them that marks the end of the string, and this is called NULL. So the first thing flashSequence does is to define a variable called i. This is going to indicate the current position in the string of dots and dashes, starting at position 0. The while loop will keep going until we reach the NULL on the end of the string.

Inside the while loop, we first flash the current dot or dash using a function that we are going to discuss in a moment and then add 1 to i and go back round the loop flashing each dot or dash in turn until we reach the end of the string.

The final function that we have defined is flashDotOrDash; this just turns the LED on and then uses an if statement to either delay for the duration of a single dot if the character is a dot or for three times that duration if the character is a dash before it turns the LED off again.

Putting It All Together

Load the completed sketch for Project 3 from your Arduino Sketchbook and download it onto your board (see Chapter 1).

To use the Morse code translator, we need to use a part of the Arduino software called the *Serial Monitor*. This window allows you to type messages that are sent to the Arduino board as well as see any messages that the Arduino board chooses to reply with.

The Serial Monitor is launched by clicking the rightmost icon shown highlighted in Figure 3-2.

The Serial Monitor (see Figure 3-3) has two parts. At the top, there is a field into which a line of text can be typed that will be sent to the board when you either click Send or press RETURN.

Below that is a larger area in which any messages coming from the Arduino board will be displayed. Right at the bottom of the window is a drop-down list where you can select the speed at which the data is sent. Whatever you select here must match the baud rate that you specify in your script's startup message. We use 9600, which is the default, so there is no need to change anything here.

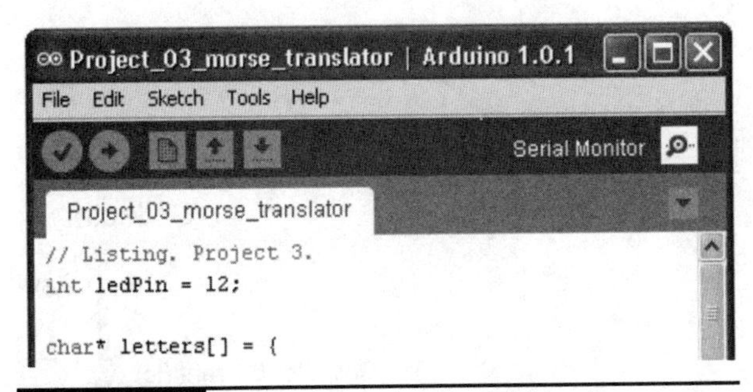

Figure 3-2 Launching the Serial Monitor.

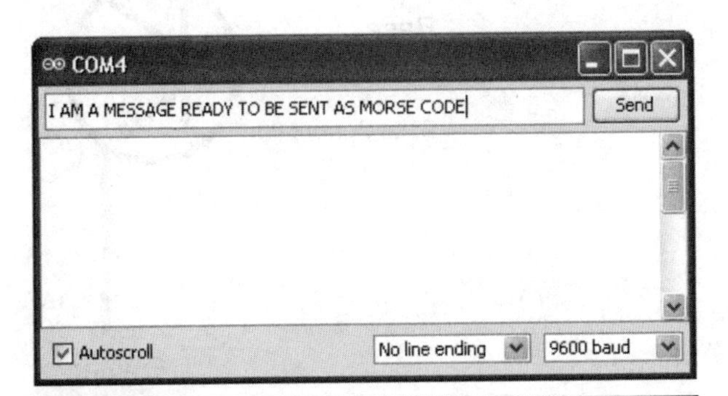

Figure 3-3 The Serial Monitor window.

So all we need to do is launch the Serial Monitor, type some text into the Send field, and click the Send button or press RETURN. We should then have our message flashed to us in Morse code.

Project 4
High-Brightness Morse Code Translator

The little LED on Project 3 is unlikely to be visible from the ship on the horizon being lured by our bogus Evil Genius distress message. So, in this project, we are going to up the power and use a 1 W Luxeon LED. These LEDs are extremely bright, and all the light comes from a tiny little area in the center, so to avoid any possibility of retina damage, do not stare directly into it.

We also look at how, with a bit of soldering, we can make this project into a shield that can be plugged into our Arduino board.

COMPONENTS AND EQUIPMENT

	Description	Appendix
	Arduino Uno or Leonardo	m1/m2
D1	Luxeon 1 W LED	s10
R1	270 Ω, 0.25 W resistor	r3
R2	4.7 Ω, 0.25 W resistor	r1
T1	BD139 power transistor	s17
	Solderless beadboard	h1
	Jumper wires	h2
	Protoshield kit (optional)	m4

Hardware

The LED we used in Project 3 used about 10 mA at 2V. We can use this to calculate power using the formula

$$P = I V$$

Power equals the voltage across something times the current flowing through it, and the unit of power is the watt. So that LED would be approximately 20 mW, or a fiftieth of the power of our 1 W Luxeon LED. While an Arduino will cope just fine driving a 20 mW LED, it will not be able to directly drive the 1 W LED.

This is a common problem in electronics and can be summed up as getting a small current to control a bigger current, something that is known as *amplification*. The most commonly used electronic component for amplification is the transistor, so that is what we will use to switch our Luxeon LED on and off.

The basic operation of a transistor is shown in Figure 3-4. There are many different types of transistors, and probably the most common and the type that we are going to use is called an *NPN bipolar transistor*.

This transistor has three leads: the emitter, the collector, and the base. And the basic principle is that a small current flowing through the base will allow a much bigger current to flow between the collector and the emitter.

Just how much bigger the current is depends on the transistor, but it is typically a factor of 100. So a current of 10 mA flowing through the base could cause up to 1 A to flow through the collector. So, if we kept the 270 Ω resistor that we used to drive the LED at 10 mA, we could expect it to be more than enough to allow the transistor to switch the hundreds of milliamps needed by the Luxeon LED.

The schematic diagram for our control circuit is shown in Figure 3-5.

The 270 Ω resistor (R1) limits the current that flows through the base. We can calculate the current using the formula I = V/R. V will be 4.4V

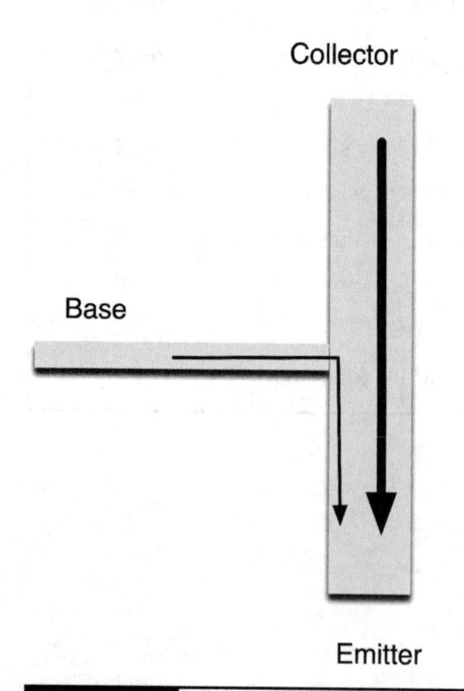

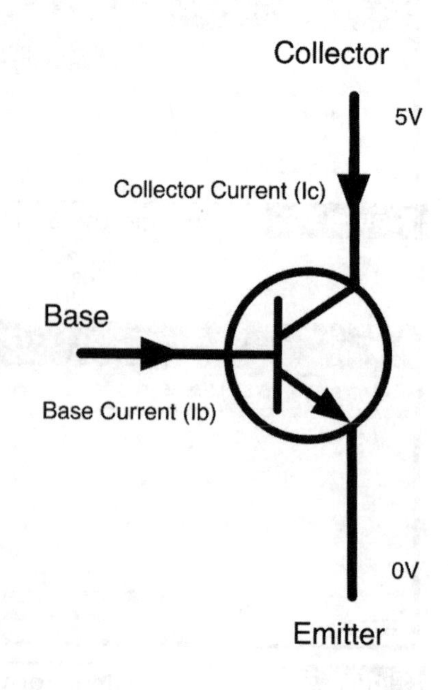

Figure 3-4 The operation of an NPN bipolar transistor.

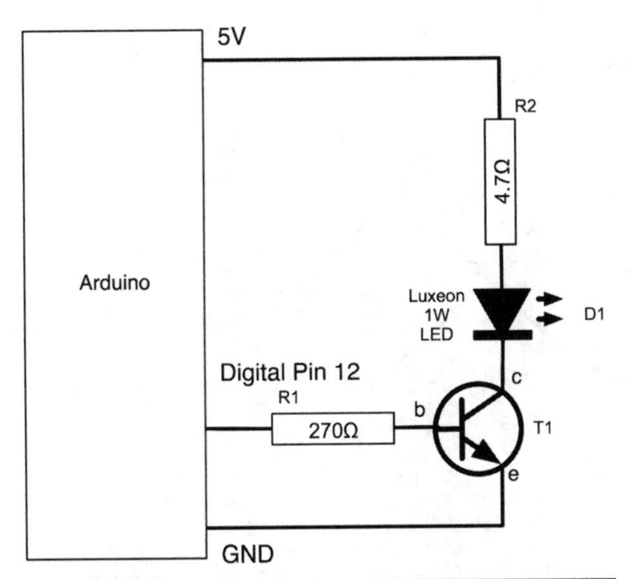

Figure 3-5 Schematic diagram for high-power LED driving.

rather than 5V because transistors normally have a voltage of 0.6V between the base and emitter, and the highest voltage the Arduino can supply from an output pin is 5V. So the current will be 4.4/270 = 16 mA.

The datasheet for this LED states that the absolute maximum forward current is 350 mA, and the forward voltage is 3.4V. So we will aim for around 200 mA, which will make the LED good and bright without shortening its life.

R2 limits the current flowing through the LED to around 200 mA. We came up with the figure of 4.7 Ω by using the formula R = V/I. V will be roughly 5 – 3.4 – 0.6 = 1.0V. 5V is the supply voltage, the LED drops roughly 3.4V and the transistor 0.6V, so the resistance should be 1.0V/200 mA = 5 Ω. Resistors come in standard values, so we will select a 4.7 Ω resistor. We must also use a resistor that can cope with this relatively high current. The power that the resistor will burn off as heat is equal to the voltage across it multiplied by the current flowing through it. In this case, that is 200 mA × 1.0V, which is 200 mW. This means that a regular 0.5 W or even 0.25 W resistor will be just fine.

In the same way, when choosing a transistor, we need to make sure that it can handle the power. When it is turned on, the transistor will consume power equal to current times voltage. In this case, the base current is small enough to ignore, so the power will just be 0.6V × 200 mA, or 120 mW. It is always a good idea to pick a transistor that can easily cope with the power. In this case, we are going to use a BD139, which has a power rating of over 12 W. In Chapter 10 you can find a table of commonly used transistors.

Now we need to put our components into the breadboard according to the layout shown in Figure 3-6, with the corresponding photograph of Figure 3-8. It is crucial to correctly identify the leads of the transistor and the LED. The metallic side of the transistor should be facing the board. The LED will have a little + symbol next to the positive connection.

Later in this project we are going to show you how you can move the project from the breadboard to a more permanent design using the Arduino Protoshield. This requires some soldering, so if you think you might go on to make a shield and have the facilities to solder, I would solder some leads onto the Luxeon LED. Solder short lengths of solid-core wire to two of the six tags around the edge. They should be marked + and –. It is a good idea to color-code your leads with red for positive and blue or black for negative.

If you do not want to solder, that's fine; you just need to carefully twist the solid-core wire around the connectors as shown in Figure 3-7.

Figure 3-8 shows the fully assembled breadboard.

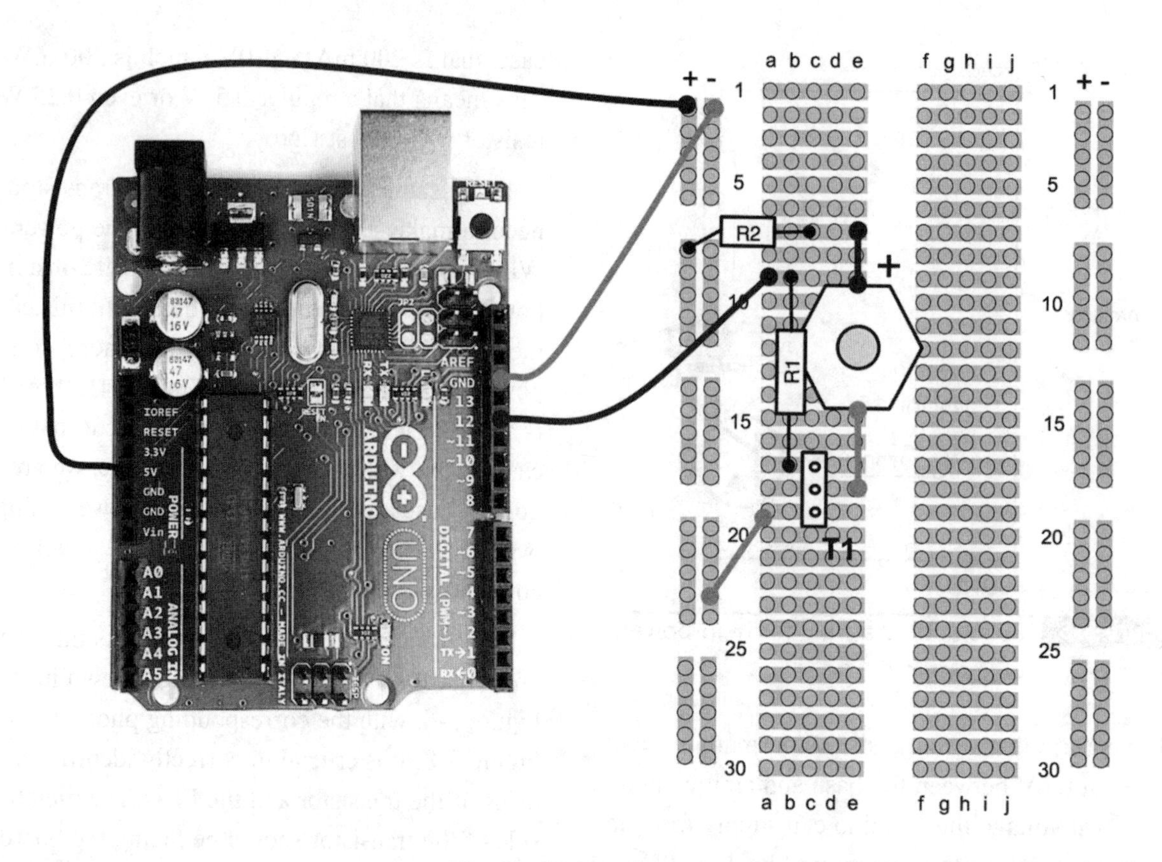

T1 - Metal face to the left, Triangular pattern to the right.

Figure 3-6 Project 4 breadboard layout.

Figure 3-7 Attaching leads to the Luxeon LED without soldering.

Software

Project 4 uses exactly the same sketch as Project 3.

Putting It All Together

If you do not still have the sketch from Project 3 loaded, then load the sketch for Project 3 from your Arduino Sketchbook and download it onto your board (see Chapter 1).

Again, testing the project is the same as for Project 3. You will need to open the Serial Monitor window and just start typing.

The LED actually has a very wide angle of view, so one variation on this project would be to adapt an LED torch in which the LED has a reflector to focus the beam.

Making a Shield

This is the first project that we have made that has enough components to justify making an Arduino Shield circuit board to sit on top of the Arduino board itself. We are also going to use this hardware with minor modifications in Project 6, so perhaps it is time to make ourselves a Luxeon LED Shield.

Figure 3-8 Photograph of complete breadboard for Project 4.

Making your own circuit boards at home is perfectly possible but requires the use of noxious chemicals and a fair amount of equipment. But fortunately, there is another great piece of Arduino-related open-source hardware called the *Arduino Protoshield*. If you shop around, these can be obtained for $10 or less and will provide you with a kit of all you need to make a basic shield. That includes the board itself, the header connector pins that fit into the Arduino, and some LEDs, switches, and resistors. Please be aware that there are several variations of the Protoshield board, so you may have to adapt the following design if your board is slightly different.

The components for a Protoshield are shown in Figure 3-9, the most important part being the Protoshield circuit board (PCB). It is possible to just buy the PCB on its own, which for many projects will be all you need.

We are not going to solder all the components that came with our kit onto the board. We are just going to add the power LED, its resistor, and just the bottom pins that connect to the Arduino board

Figure 3-9 Protoshield in kit form.

because this is going to be a top shield and will not have any other shields on top of it.

A good guide for assembling circuit boards is to solder in place the lowest components first. So in this case we will solder the resistors, the LED, the reset switch, and then the bottom pin connectors.

The 1K resistor, LED, and switch are all pushed through from the top of the board and soldered underneath (Figure 3-10). The short part of the connector pins will be pushed up from underneath the board and soldered on top.

When soldering the connector pins, make sure that they are lined up correctly because there are two parallel rows for the connectors: one for the connection to the pins below and one for the sockets, which we are not using, that are intended to connect to further shields.

A good way to ensure that the headers are in the right place is to fit the sections of header into an Arduino board and then place the shield on top and solder the pins while it's still plugged into the Arduino board. This will also ensure that the pins are straight.

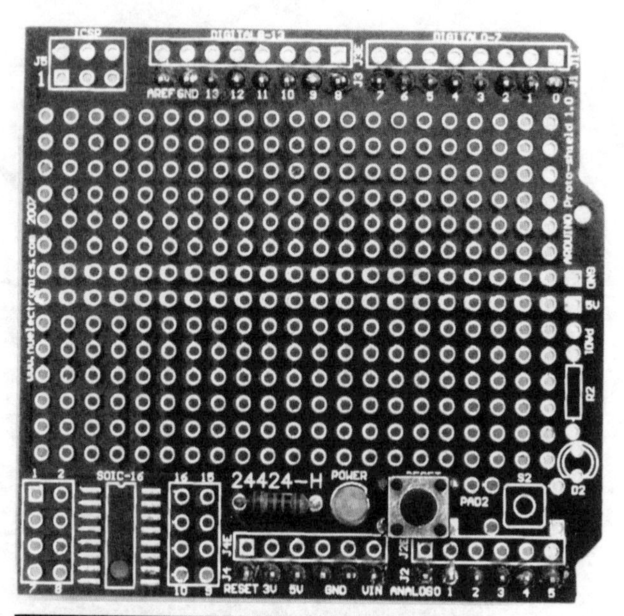

Figure 3-11 Assembled basic Protoshield.

When all the components have been soldered in place, you should have a board that looks like Figure 3-11.

We can now add our components for this project, which we can take from the breadboard. First, line up all the components in their intended places according to the layout of Figure 3-12 to

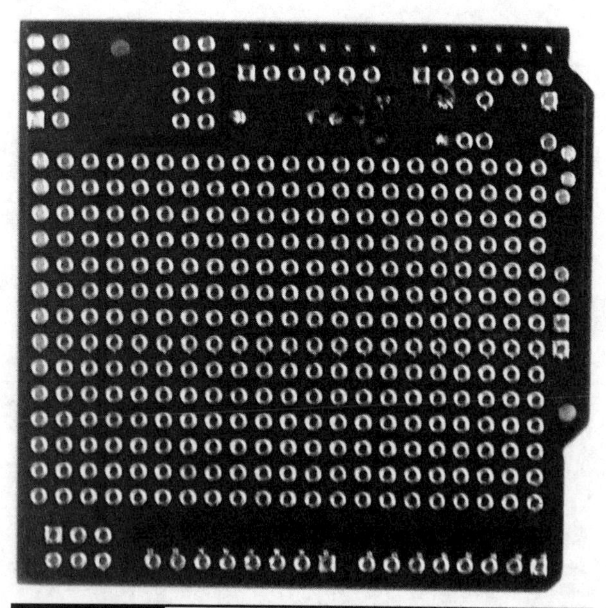

Figure 3-10 The underside of the Protoshield.

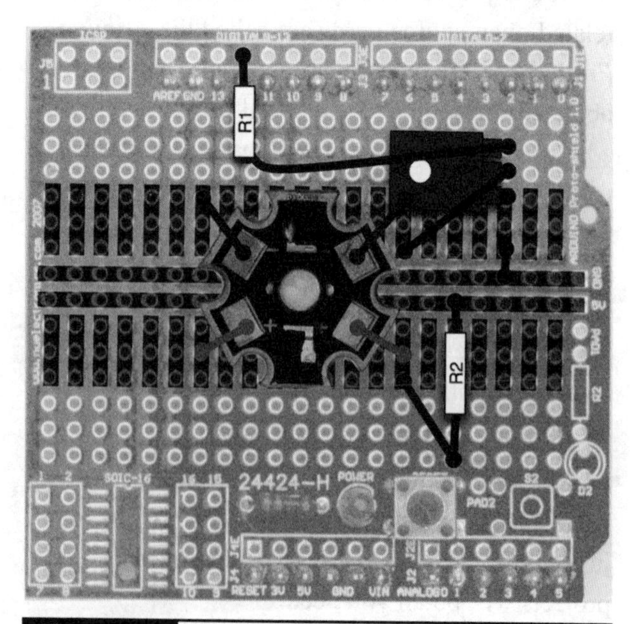

Figure 3-12 Project 4 Protoshield layout.

make sure that everything fits in the available space.

This kind of board is double-sided—that is, you can solder to the top or bottom of the board. As you can see from the layout in Figure 3-12, some of the connections are in strips like a breadboard.

We are going to mount all the components on the top side, with the leads pushed through and soldered on the underside where they emerge from the board. The leads of the components underneath can then be connected up and excess leads snipped off. If necessary, lengths of solid-core wire can be used where the leads will not reach.

Figure 3-13 shows the completed shield. Power up your board and test it out. If it does not work as soon as you power it up, disconnect it from the power right away and carefully check the shield for any short circuits or broken connections using a multimeter.

Congratulations! You have created your first Arduino Shield, and it is one that we can reuse in later projects.

Summary

So we have made a start on some simple LED projects and discovered how to use high-power Luxeon LEDs. We have also learned a bit more about programming our Arduino board in C.

In Chapter 4 we are going to extend this by looking at some more LED-based projects, including a model traffic signal and a high-power strobe light.

Figure 3-13 Complete Luxeon shield attached to an Arduino board.

CHAPTER 4

More LED Projects

In this chapter we are going to build on those versatile little components, LEDs, and learn a bit more about digital inputs and outputs, including how to use push-button switches.

The projects that we are going to build in this chapter are a model traffic signal, two strobe light projects, and a bright-light module using high-power Luxeon LEDs.

Digital Inputs and Outputs

The digital pins 0 to 12 can all be used as either an input or an output. This is set in your sketch. Since you are going to be connecting electronics to one of these pins, it is unlikely that you are going to want to change the mode of a pin. That is, once a pin is set to be an output, you are not going to change it to be an input midway through a sketch.

For this reason, it is a convention to set the direction of a digital pin in the setup function that must be defined in every sketch.

For example, the following code sets digital pin 10 to be an output and digital pin 11 to be an input. Note how we use a variable declaration in our sketch to make it easier to change the pin used for a particular purpose later on.

In the sketch for Project 5, we will connect pin 5 to a switch that will connect it to GND when it is pressed. The pinMode of pin 5 is set to be INPUT_PULLUP rather than just INPUT. This sets the input to be "pulled up" to HIGH. Another way of thinking of this is that the input is by default HIGH unless pulled LOW.

```
int ledPin = 10;
int switchPin = 11;

void setup()
{
  pinMode(ledPin, OUTPUT);
  pinMode(switchPin, INPUT);
}
```

Project 5
Model Traffic Signal

So now that we know how to set a digital pin to be an input, we can build a project for model traffic signals using red, yellow, and green LEDs. Every time we press the button, the traffic signal will go to the next step in the sequence. In the United Kingdom, the sequence of such traffic signals is red, red and amber together, green, amber, and then back to red.

As a bonus, if we hold the button down, the lights will change in sequence by themselves with a delay between each step.

The components for Project 5 are listed next. When using LEDs, for best effect, try to pick LEDs of similar brightness.

COMPONENTS AND EQUIPMENT		
	Description	Appendix
	Arduino Uno or Leonardo	m1/m2
D1	5-mm red LED	s1
D2	5-mm yellow LED	s3
D3	5-mm green LED	s2
R1-R3	270 Ω, 0.25 W resistor	r3
S1	Miniature push to make switch	h3
	Solderless beadboard	h1
	Jumper wires	h2

Hardware

The schematic diagram for the project is shown in Figure 4-1.

The LEDs are connected in the same way as our earlier project, each with a current-limiting resistor. Pressing the push-button switch will connect digital pin 5 to GND.

A photograph of the project is shown in Figure 4-2 and the board layout in Figure 4-3.

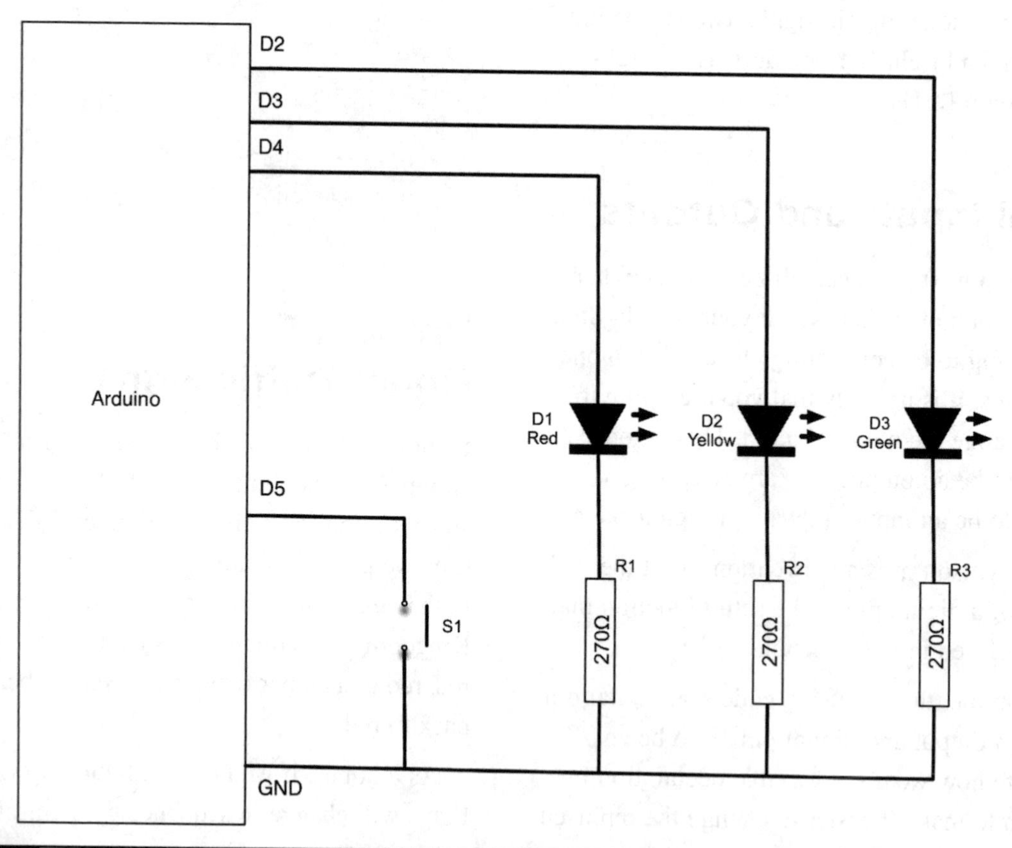

Figure 4-1 Schematic diagram for Project 5.

Figure 4-2 Project 5: a model traffic signal.

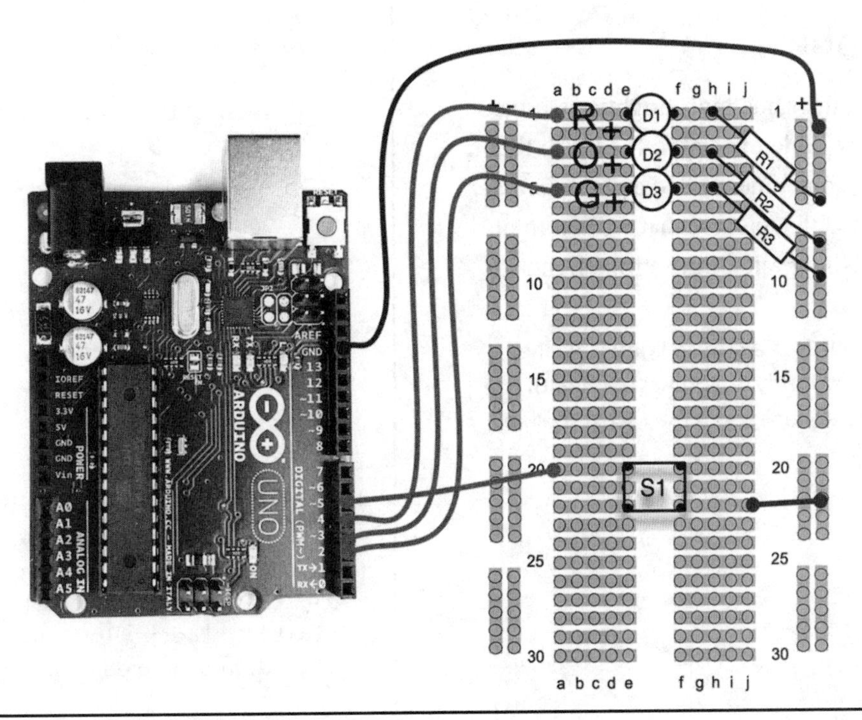

Figure 4-3 Breadboard layout for Project 5.

Software

The sketch for Project 5 is shown in Listing Project 5.

The sketch is fairly self-explanatory. We only check to see if the switch is pressed once a second so that pressing the switch rapidly will not move the light sequence on. However, if we press and hold the switch, the lights will automatically sequence round.

The delay(1000) command prevents the lights changing so fast that they are a blur.

We use a separate function setLights to set the state of each LED, reducing three lines of code to one.

Putting It All Together

Load the completed sketch for Project 5 from your Arduino Sketchbook (see Chapter 1).

Test the project by holding down the button and making sure that the LEDs all light in sequence.

LISTING PROJECT 5

```
int redPin = 4;
int yellowPin = 3;
int greenPin = 2;
int buttonPin = 5;

int state = 0;

void setup()
{
  pinMode(redPin, OUTPUT);
  pinMode(yellowPin, OUTPUT);
  pinMode(greenPin, OUTPUT);
  pinMode(buttonPin, INPUT_PULLUP);
}

void loop()
{
  if (digitalRead(buttonPin))
  {
    if (state == 0)
    {
      setLights(HIGH, LOW, LOW);
      state = 1;
    }
    else if (state == 1)
    {
      setLights(HIGH, HIGH, LOW);
      state = 2;
    }
    else if (state == 2)
    {
      setLights(LOW, LOW, HIGH);
      state = 3;
    }
    else if (state == 3)
    {
      setLights(LOW, HIGH, LOW);
      state = 0;
    }
    delay(1000);
  }
}

void setLights(int red, int yellow,
               int green)
{
  digitalWrite(redPin, red);
  digitalWrite(yellowPin, yellow);
  digitalWrite(greenPin, green);
}
```

Project 6
Strobe Light

This project uses the same high-brightness Luxeon LED as the Morse code translator. It adds to that a variable resistor, sometimes called a *potentiometer*. This provides us with a control that we can rotate to control the flashing rate of the strobe light.

CAUTION **This is a strobe light; it flashes brightly. If you have a health condition such as epilepsy, you may wish to skip this project.**

COMPONENTS AND EQUIPMENT		
	Description	Appendix
	Arduino Uno or Leonardo	m1/m2
D1	Luxeon 1 W LED	s10
R1	270 Ω, 0.25 W resistor	r3
R2	4.7 Ω, 0.25 W resistor	r1
T1	BD139 power transistor	s17
R3	10K linear potentiometer (trimpot)	r11
	Protoshield kit (optional)	m4
	2.1-mm DC power jack (optional)	h4
	9V battery clip (optional)	h5

Hardware

The hardware for this project is basically the same as for Project 4 but with the addition of a variable resistor or potentiometer as they are sometimes known (Figure 4-4).

The Arduino is equipped with six analog input pins numbered Analog 0 to Analog 5. These measure the voltage at their input and give a number between 0 (0V) and 1023 (5V).

We can use this to detect the position of a control knob by connecting a variable resistor acting as a potential divider to our analog pin. Figure 4-5 shows the internal structure of a variable resistor.

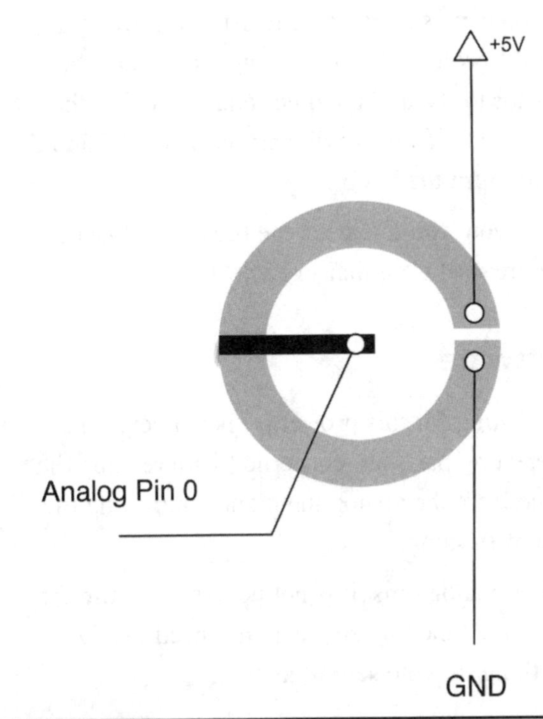

Figure 4-5 The internal workings of a variable resistor.

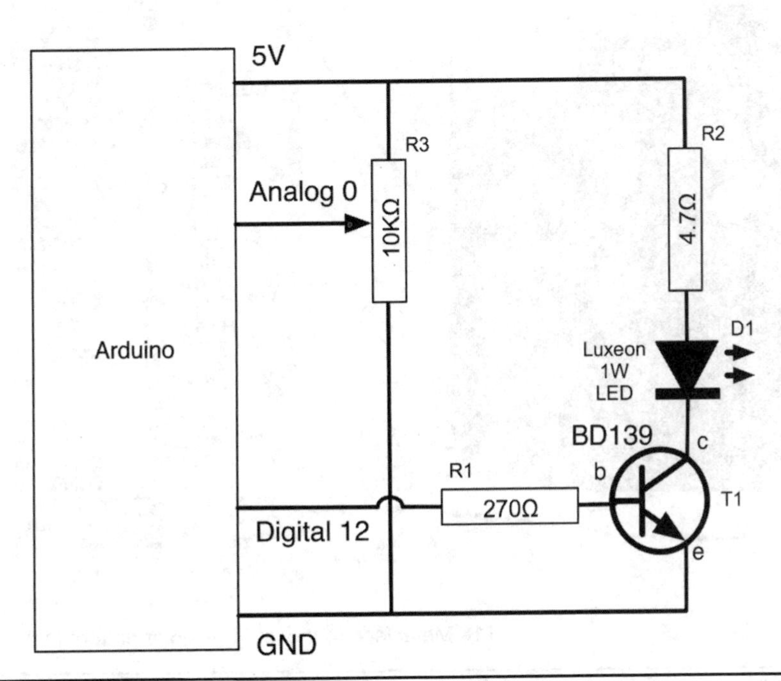

Figure 4-4 Schematic diagram for Project 6.

A variable resistor is a component that is typically used for volume control. It is constructed as a circular conductive track with a gap in it and connections at both ends. A slider provides a movable third connection.

You can use a variable resistor to provide a variable voltage by connecting one end of the resistor to 0V and the other end to 5V, and then the voltage at the slider will vary between 0V and 5V as you turn the knob.

As you would expect, the breadboard layout (Figure 4-6) is similar to Project 4.

Software

The listing for this project is shown here. The interesting parts are concerned with reading the value from the analog input and controlling the rate of flashing.

For analog pins, it is not necessary to use the pinMode function, so we do not need to add anything into the setup function.

LISTING PROJECT 6

```
int ledPin = 12;
int analogPin = 0;

void setup()
{
  pinMode(ledPin, OUTPUT);
}

void loop()
{
  int period = (1023 -
      analogRead(analogPin)) / 2 + 25;
  digitalWrite(ledPin, HIGH);
  delay(period);
  digitalWrite(ledPin, LOW);
  delay(period);
}
```

Let us say that we are going to vary the rate of flashing between once a second and 20 times a second; the delays between turning the LED on and off will be 500 milliseconds and 25 milliseconds, respectively.

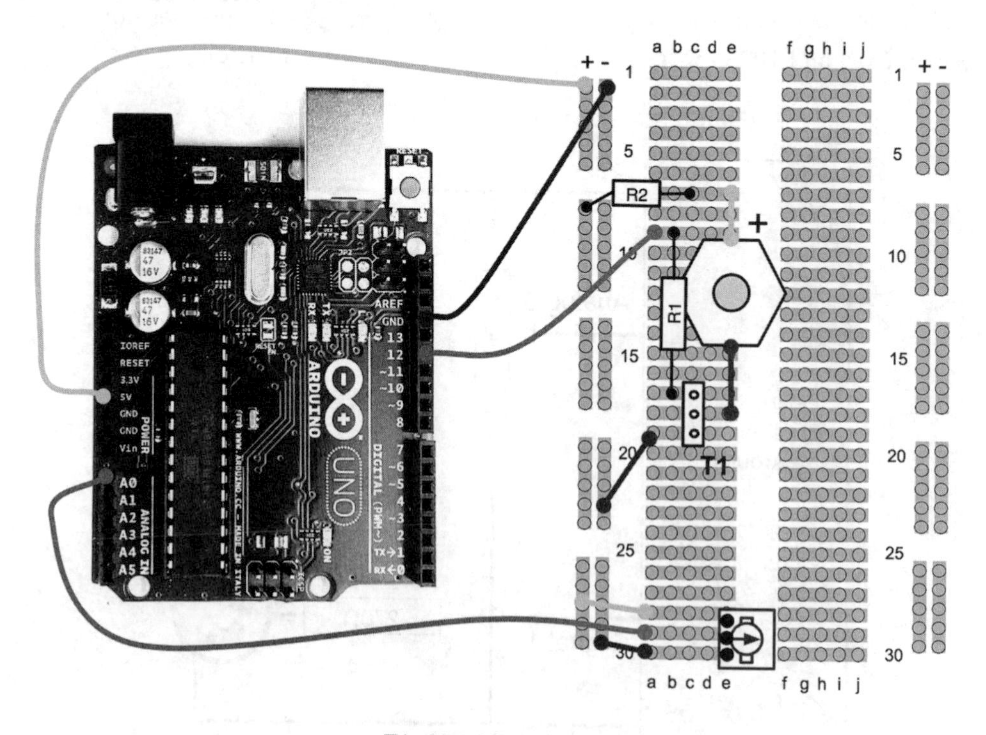

T1 - Metal face to the left, triangular pattern to the right.

Figure 4-6 Breadboard layout for Project 6.

So, if our analog input changes from 0 to 1023, the calculation that we need to determine the flash delay is roughly

```
flash_delay = (1023 - analog_value)
    / 2 + 25
```

So an analog_value of 0 would give a flash_delay of 561 and an analog_value of 1023 would give a delay of 25. We should actually be dividing by slightly more than 2, but it makes things easier if we keep everything as integers.

Putting It All Together

Load the completed sketch for Project 6 from your Arduino Sketchbook and download it to the board (see Chapter 1).

You will find that turning the variable resistor control clockwise will increase the rate of flashing as the voltage at the analog input increases. Turning it counterclockwise will slow the rate of flashing.

Making a Shield

If you want to make a shield for this project, you can either adapt the shield for Project 4 or create a new shield from scratch.

The layout of components on the Protoshield is shown in Figure 4-7.

This is basically the same as for Project 4, except that we have added the variable resistor. The pins on a variable resistor are too thick to fit into the holes on the Protoshield, so you can either attach it using wires or, as we have done here, carefully solder the leads to the top surface where they touch the board. To provide some mechanical strength, the variable resistor can be glued in place first with a drop of Super Glue. The wiring for the variable resistor to 5V, GND, and Analog 0 can be made underneath the board out of sight.

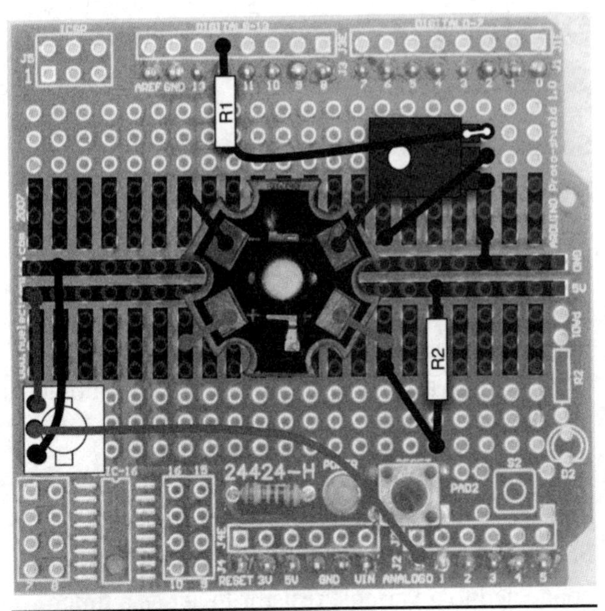

Figure 4-7 Protoshield layout for Project 6.

Having made a shield, we can make the project independent of our computer by powering it from a 9V battery.

To power the project from a battery, we need to make ourselves a small lead that has a PP3 battery clip on one end and a 2.1-mm power plug on the other. Figure 4-8 shows the semiassembled lead. You can also buy such leads ready assembled from Sparkfun and Adafruit.

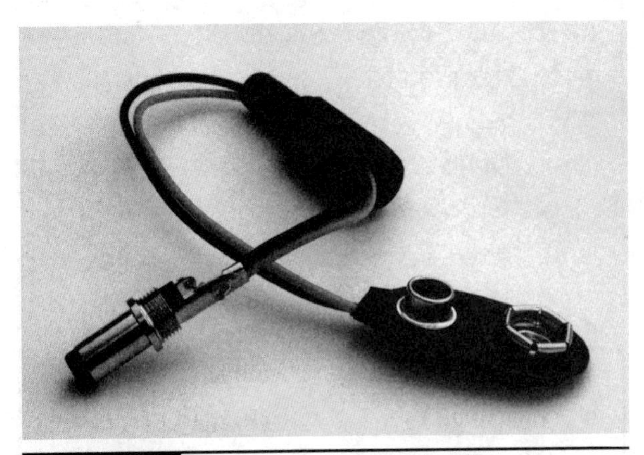

Figure 4-8 Creating a battery lead.

Project 7
SAD Light

Seasonal affective disorder (SAD) affects a great number of people, and research has shown that exposure to a bright white light that mimics daylight for 10 or 20 minutes a day has a beneficial effect. To use this project for such a purpose, I would suggest the use of some kind of diffuser such as frosted glass because you should not stare directly at the point light sources of the LEDs.

This is another project based on Luxeon high-brightness LEDs. We will use an analog input connected to a variable resistor to act as a timer control, turning the LED on for a given period set by the position of the variable resistor's slider. We will also use an analog output to slowly raise the brightness of the LEDs as they turn on and then slowly decrease it as they turn off. To make the light bright enough to be of use as a SAD light, we are going to use not just one Luxeon LED but six.

At this point the caring nature of this project may be causing the Evil Genius something of an identity crisis. But fear not—in Project 8 we will turn this same hardware into a fearsome high-powered strobe light.

COMPONENTS AND EQUIPMENT

	Description	Appendix
	Arduino Uno or Leonardo	m1/m2
D1-6	Luxeon 1 W LED	s10
R1-3	1 kΩ < 0.25 W resistor	r5
R4-5	4.7 Ω < 0.25 W resistor	r1
R6	100K linear potentiometer	r12
IC1-2	LM317 voltage regulator	s18
T1-2	2N7000 FET	s15
	Regulated 15V 1 A power supply	h8
	Perf board	h9
	Three-way screw terminal	h10

- Please note that this is one of the projects in this book that requires soldering.
- You are going to need six Luxeon LEDs for this project. If you want to save some money, look at online auctions, where ten of these should be available for $10 to $20.

Hardware

Some of the digital pins, namely, digital pins 5, 6, 9, 10, and 11, on an Uno and a few more on a Leonardo can provide a variable output rather than just 5V or nothing. These are the pins with "PWM" next to them on the board.

PWM stands for *pulse-width modulation* and refers to the means of controlling the amount of power at the output. It does so by rapidly turning the output on and off.

The pulses are always delivered at the same rate (roughly 500 per second), but the length of the pulses is varied. If the pulse is long, our LED will be on all the time. If, however, the pulses are short, the LED is only actually lit for a small portion of the time. This happens too fast for the observer to even tell that the LED is flickering, and it just appears that the LED is brighter or dimmer.

You will meet PWM again in Project 19, where we use it to generate sounds.

The value of the output can be set using the function analogWrite, which requires an output value between 0 and 255, where 0 will be off and 255 will be full power.

As you can see from the schematic diagram in Figure 4-9, the LEDs are arranged in two columns of three. The LEDs are also supplied from an external 15V supply rather than the 5V supply that we used before. Since each LED consumes about 300 mA, each column will draw about 300 mA, so the supply must be capable of supplying 0.6 A (1 A to be on the safe side).

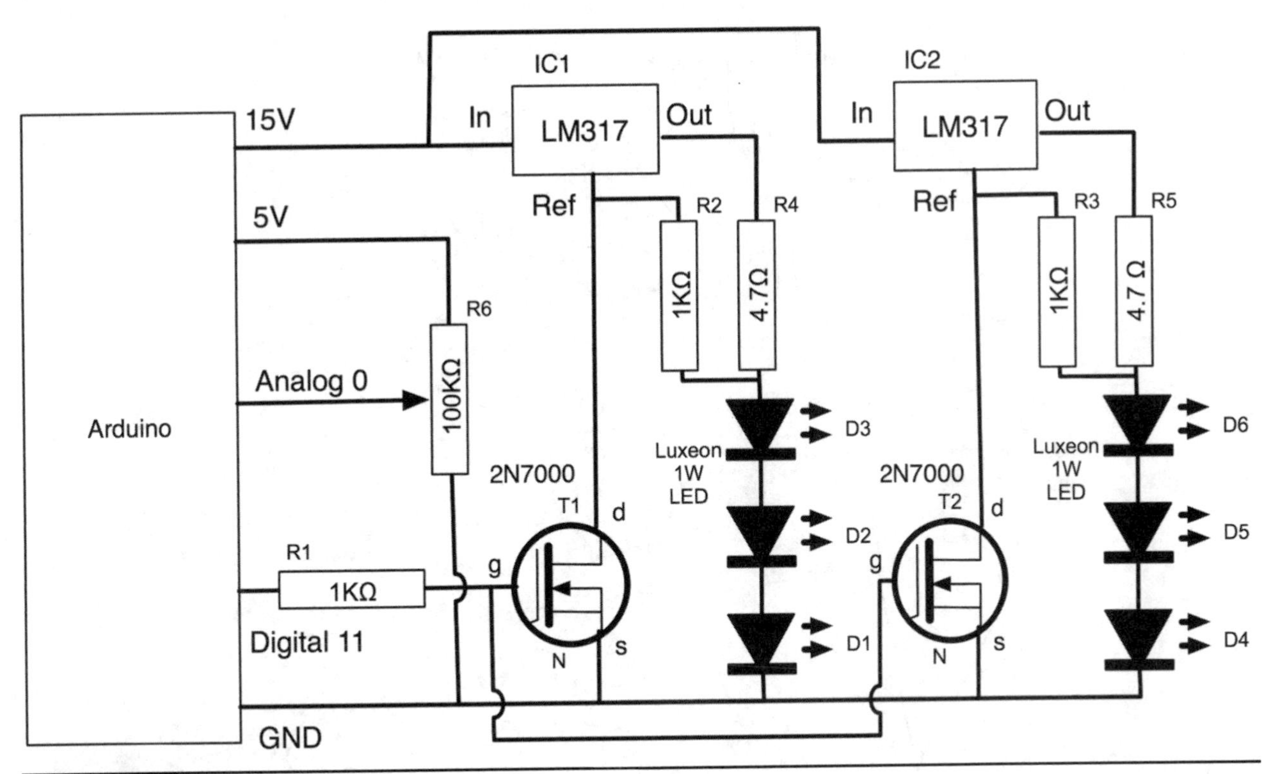

Figure 4-9 Schematic diagram for Project 7.

This is the most complex schematic so far in our projects. We are using two integrated-circuit variable voltage regulators to limit the current flowing to the LEDs. The output of the voltage regulators will normally be 1.25V above whatever the voltage is at the Ref pin of the chip. This means that if we drive our LEDs through a 4 W resistor, there will be a current of roughly I = V/R, or 1.25/4 = 312 mA flowing through it (which is about right).

The field-effect transistor (FET) is like our normal bipolar transistor in that it can act as a switch, but it has a very high off resistance. So, when it is not triggered by a voltage at its gate, it's as if it isn't there in the circuit. However, when it is turned on, it will pull down the voltage at the regulator's Ref pin to a low enough voltage to prevent any current flowing into the LEDs, turning them off. Both the FETs are controlled from the same digital pin 11.

The completed LED module is shown in Figure 4-10 and the perf board layout in Figure 4-11.

Figure 4-10 Project 7: high-power light module.

The module is built on perf (perforated) board. The perf board is just a board with holes in it. It has no connections at all. So it acts as a structure on which to fit your components, but you have to wire them up on the underside of the board, either by connecting their leads together or by adding wires.

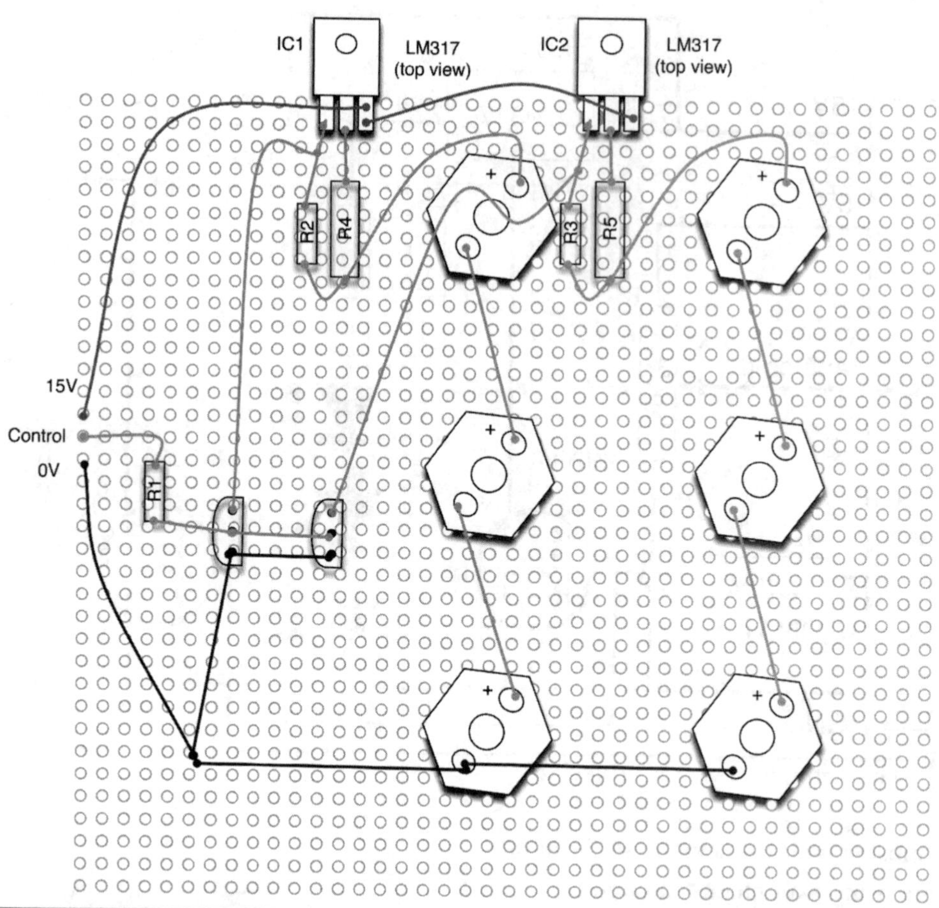

Figure 4-11 Perf board layout.

It is easier to solder two wires onto each LED before fitting it onto the board. It is a good idea to color-code those leads—red for positive and black or blue for negative—so that you get the LEDs in the correct way round.

The LEDs will get hot, so it is a good idea to leave a gap between them and the perf board using the insulation on the wire to act as a spacer. The voltage regulator will also get hot but should be okay without a heat sink. The voltage regulator integrator circuits (ICs) actually have built-in thermal protection and will automatically reduce the current if they start to get too hot.

The screw terminals on the board are for the power supply GND and 15V and a control input. When we connect this to the Arduino board, the 15V will come from the Vin pin on the Arduino, which in turn is supplied from a 15V power supply.

Our high-power LED module will be of use in other projects, so we are going to plug the variable resistor directly into the "Analog In" strip of connectors on the Arduino board. The spacing of pins on the variable resistor is 1/5 of an inch, which means that if the middle slider pin is in the socket for Analog 2, the other two pins will be in the sockets for Analog 0 and Analog 4. You can see this arrangement in Figure 4-12.

So, in order to have 5V at one end of our variable resistor and 0V at the other, we are going to set the outputs of analog pins 0 and to 0V and 5V, respectively.

Software

At the top of the sketch, after the variable used for pins, we have four other variables: startupSeconds, turnOffSeconds, minOnSeconds, and

LISTING PROJECT 7

```
int ledPin = 11;
int analogPin = A2;

int startupSeconds = 20;
int turnOffSeconds = 10;
int minOnSeconds = 300;
int maxOnSeconds = 1800;

int brightness = 0;

void setup()
{
  pinMode(ledPin, OUTPUT);
  digitalWrite(ledPin, HIGH);
  pinMode(A0, OUTPUT);                    // Use Analog pins 0 and 4 for
  pinMode(A4, OUTPUT);                    // the variable resistor
  digitalWrite(A4, HIGH);
  int analogIn = analogRead(analogPin);
  int onTime = map(analogIn, 0, 1023, minOnSeconds, maxOnSeconds);
  turnOn();
  delay(onTime * 1000);
  turnOff();
}

void turnOn()
{
  brightness = 0;
  int period = startupSeconds * 1000 / 256;
  while (brightness < 255)
  {
    analogWrite(ledPin, 255 - brightness);
    delay(period);
    brightness ++;
  }
}

void turnOff()
{
  int period = turnOffSeconds * 1000 / 256;
  while (brightness >= 0)
  {
    analogWrite(ledPin, 255 - brightness);
    delay(period);
    brightness -;
  }
}

void loop()
{}
```

maxOnSeconds. This is common practice in programming. By putting these values that we might want to change into variables and making them visible at the top of the sketch, it makes it easier to change them.

The variable startupSeconds determines how long it will take for the brightness of the LEDs to be gradually raised until it reaches maximum brightness. Similarly, turnOffSeconds determines the time period for dimming the LEDs. The variables minOnSeconds and maxOnSeconds determine the range of times set by the variable resistor.

In this sketch, there is nothing in the loop function. Instead, all the code is in setup. So the light will automatically start its cycle when it is powered up. Once it has finished, it will stay turned off until the Reset button is pressed.

The slow turn-on is accomplished by gradually increasing the value of the analog output by 1. This is carried out in a while loop, where the delay is set to 1/255 of the startup time so that after 255 steps maximum brightness has been achieved. Slow turn-off works in a similar manner.

The time period at full brightness is set by the analog input. Assuming that we want a range of times from 5 to 30 minutes, we need to convert the value of 0 to 1023 to a number of seconds between 300 and 1800. Fortunately, there is a handy Arduino function that we can use to do this. The function map takes five arguments: the value you want to convert, the minimum input value (0 in this case), the maximum input value (1023), the minimum output value (300), and the maximum output value (1800).

Putting It All Together

Load the completed sketch for Project 7 from your Arduino Sketchbook and download it to the board (see Chapter 1).

You now need to attach wires from the Vin, GND, and digital pin 11 of the Arduino board to the three screw terminals of the LED module (Figure 4-12). Plug a 15V power supply into the board's power socket, and you are ready to try it.

To start the light sequence again, click the Reset button.

Figure 4-12 Project 7: SAD light.

Project 8
High-Powered Strobe Light

For this project you can use the six Luxeon LED module of Project 7 or you can use the Luxeon shield that we created for Project 4. The software will be almost the same in both cases.

In this version of the strobe light, we are going to control the strobe light effect from the computer with commands. We will send the following commands over the USB connection using the Serial Monitor:

0-9	Sets the speed of the following mode commands: 0 for off, 1 for slow, and 9 for fast
w	Wave effect gradually getting lighter then darker
s	Strobe effect

Hardware

See Project 4 (the Morse code translator using a single Luxeon LED shield) or Project 7 (array of six Luxeon LEDs) for components and construction details. Note that if you chose to reuse Project 7, you will need to change ledPin in the sketch below to use pin 11 rather than pin 12.

Software

This sketch uses the sin function to produce a nice, gently increasing brightness effect. Apart from that, the techniques we use in this sketch have mostly been used in earlier projects.

LISTING PROJECT 8

```
int ledPin = 12;

int period = 100;

char mode = 'o'; // o-off, s-strobe, w-wave

void setup()
{
  pinMode(ledPin, OUTPUT);
  analogWrite(ledPin, 255);
  Serial.begin(9600);
}

void loop()
{
  if (Serial.available())
  {
    char ch = Serial.read();
    if (ch == '0')
    {
      mode = 0;
      analogWrite(ledPin, 255);
```

(continued on next page)

LISTING PROJECT 8 (*continued*)

```
    }
    else if (ch > '0' && ch <= '9')
    {
      setPeriod(ch);
    }
    else if (ch == 'w' || ch == 's')
    {
      mode = ch;
    }
  }
  if (mode == 'w')
  {
    waveLoop();
  }
  else if (mode == 's')
  {
    strobeLoop();
  }
}

void setPeriod(char ch)
{
  int period1to9 = 9 - (ch - '0');
  period = map(period1to9, 0, 9, 50, 500);
}

void waveLoop()
{
  static float angle = 0.0;
  angle = angle + 0.01;
  if (angle > 3.142)
  {
    angle = 0;
  }
  // analogWrite(ledPin, 255 - (int)255 * sin(angle)); // Breadboard
  analogWrite(ledPin, (int)255 * sin(angle));          // Shield
  delay(period / 100);
}

void strobeLoop()
{
  //analogWrite(ledPin, 0);        // breadboard
  analogWrite(ledPin, 255);        // shield
  delay(10);
  //analogWrite(ledPin, 255);      // breadboard
  analogWrite(ledPin, 0);          // shield
  delay(period);
}
```

Putting It All Together

Load the completed sketch for Project 8 from your Arduino Sketchbook and download it to the board (see Chapter 1).

When you have installed the sketch and fitted the Luxeon shield or connected the bright six Luxeon panel, initially the lights will be off. Open the Serial Monitor window, type **s**, and press RETURN. This will start the light flashing. Try the speed commands 1 to 9. Then try typing the **w** command to switch to wave mode.

Random Number Generation

Computers are deterministic. If you ask them the same question twice, you should get the same answer. However, sometimes you want a chance to take a hand. This is obviously useful for games.

It is also useful in other circumstances—for example, a "random walk," where a robot makes a random turn, then moves forward a random distance or until it hits something, and then reverses and turns again, is much better at ensuring that the robot covers the whole area of a room than a more fixed algorithm that can result in the robot getting stuck in a pattern.

The Arduino library includes a function for creating random numbers.

There are two flavors of the function random. It can either take two arguments (minimum and maximum) or one argument (maximum), in which case the minimum is assumed to be 0.

Beware, though, because the maximum argument is misleading because the highest number you can actually get back is the maximum minus one.

So the following line will give *x* a value between 1 and 6:

```
int x = random(1, 7);
```

and the following line will give *x* a value between 0 and 9:

```
int x = random(10);
```

As we pointed out at the start of this section, computers are deterministic, and actually our random numbers are not random at all, but a long sequence of numbers with a random distribution. You will actually get the same sequence of numbers every time you run your script.

A second function (randomSeed) allows you to control this. The randomSeed function determines where in its sequence of pseudorandom numbers the random number generator starts.

A good trick is to use the value of a disconnected analog input because this will float around at a different value and give at least 1000 different starting points for our random sequence. This wouldn't do for the lottery but is acceptable for most applications. Truly random numbers are very hard to come by and involve special hardware.

Project 9
LED Dice

This project uses what we have just learned about random numbers to create electronic dice with six LEDs and a button. Every time you press the button, the LEDs "roll" for a while before settling on a value and then flashing it.

COMPONENTS AND EQUIPMENT		
	Description	**Appendix**
	Arduino Uno or Leonardo	m1/m2
D1-7	Standard LEDs, any color	s1-s6
R1-7	270 Ω, 0.25 W resistor	r3
S1	Miniature push-to-make switch	h3
	Solderless beadboard	h1
	Jumper wires	h2

Hardware

The schematic diagram for Project 9 is shown in Figure 4-13. Each LED is driven by a separate digital output via a current-limiting resistor. The only other components are the switch and its associated pull-down resistor. All the resistors and LEDs are the same, so they are not labeled separately.

Even though a die can only have a maximum of six dots, we still need seven LEDs to have the normal arrangement of a dot in the middle for odd-numbered rolls.

Figure 4-14 shows the breadboard layout and Figure 4-15 the finished breadboard.

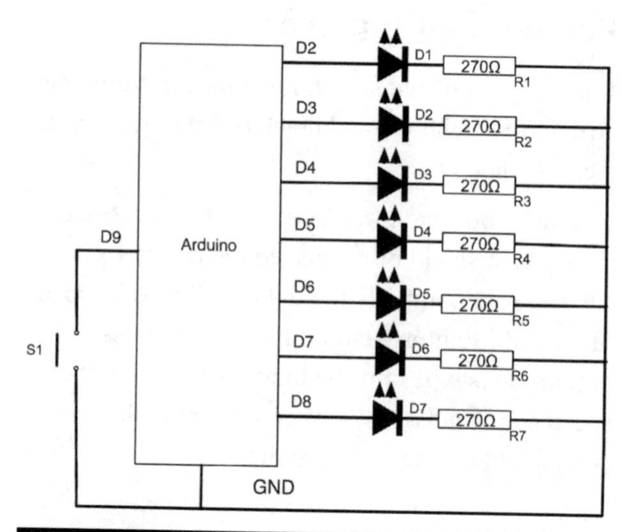

Figure 4-13 Schematic diagram for Project 9.

Software

This sketch is fairly straightforward; there are a few nice touches that make the dice behave in a similar way to real dice. For example, as the dice rolls, the number changes but gradually slows. Also, the length of time that the dice rolls is also random.

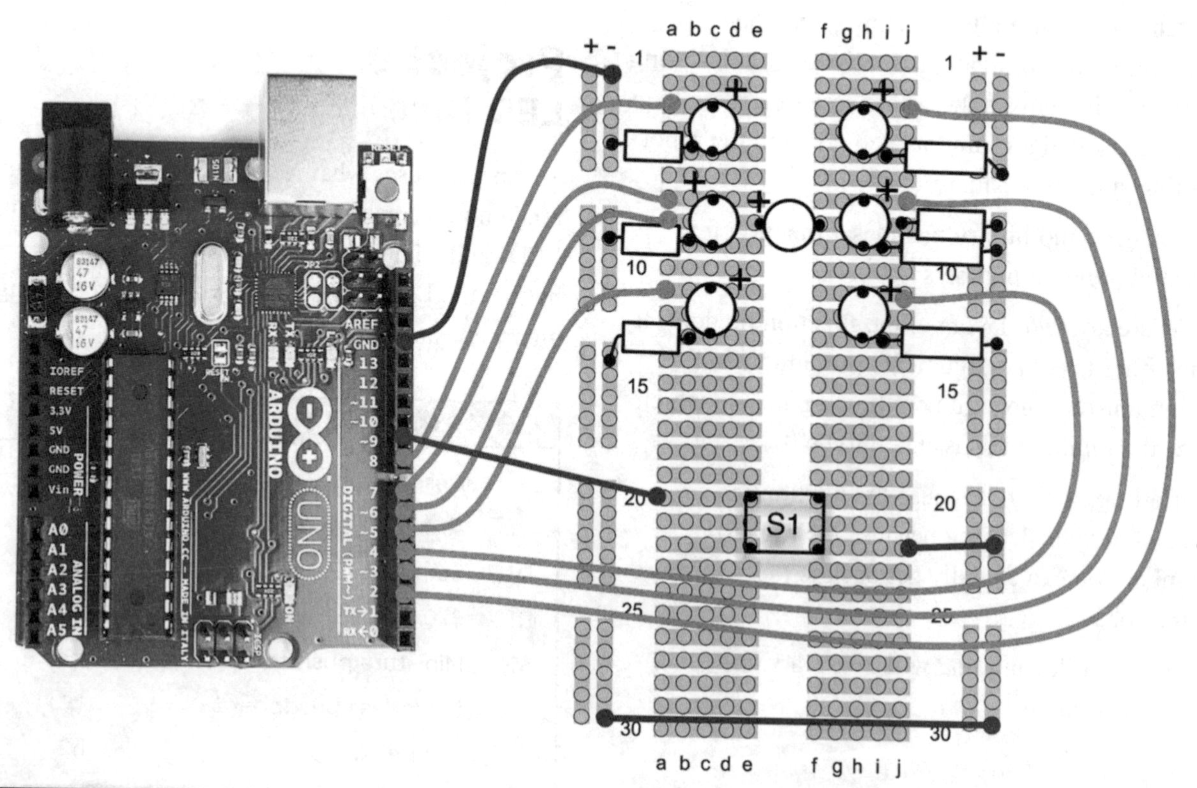

Figure 4-14 Breadboard layout for Project 9.

Figure 4-15 Project 9: LED dice.

LISTING PROJECT 9

```
int ledPins[7] = { 2, 3, 4, 5, 7, 8, 6 };
int dicePatterns[7][7] = {
  {0, 0, 0, 0, 0, 0, 1},          // 1
  {0, 0, 1, 1, 0, 0, 0},          // 2
  {0, 0, 1, 1, 0, 0, 1},          // 3
  {1, 0, 1, 1, 0, 1, 0},          // 4
  {1, 0, 1, 1, 0, 1, 1},          // 5
  {1, 1, 1, 1, 1, 1, 0},          // 6
  {0, 0, 0, 0, 0, 0, 0}           // BLANK
};

int switchPin = 9;
int blank = 6;

void setup()
{
  for (int i = 0; i < 7; i++)
  {
    pinMode(ledPins[i], OUTPUT);
```

(continued on next page)

LISTING PROJECT 9 *(continued)*

```
    digitalWrite(ledPins[i], LOW);
  }
  pinMode(switchPin, INPUT_PULLUP);
  randomSeed(analogRead(0));
}


void loop()
{
  if (digitalRead(switchPin))
  {
    rollTheDice();
  }
  delay(100);
}

void rollTheDice()
{
  int result = 0;
  int lengthOfRoll = random(15, 25);
  for (int i = 0; i < lengthOfRoll; i++)
  {
    result = random(0, 6);      // result will be 0 to 5 not 1 to 6
    show(result);
    delay(50 + i * 10);
  }
  for (int j = 0; j < 3; j++)
  {
    show(blank);
    delay(500);
    show(result);
    delay(500);
  }
}

void show(int result)
{

  for (int i = 0; i < 7; i++)
  {
  digitalWrite(ledPins[i], dicePatterns[result][i]);
  }
}
```

We now have seven LEDs to initialize in the setup method, so it is worth putting them in an array and looping over the array to initialize each pin. We also have a call to randomSeed in the setup method. If this was not there, every time we reset the board, we would end up with the same sequence of dice throws. As an experiment, you may wish to try commenting out this line by placing a // in front of it and verifying this. In fact, as an Evil Genius, you may like to omit that line so that you can cheat at *Snakes and Ladders*!

The dicePatterns array determines which LEDs should be on or off for any particular throw. So each throw element of the array is actually itself an array of seven elements, each one being either HIGH or LOW (1 or 0). When we come to display a particular result of throwing the dice, we can just loop over the array for the throw, setting each LED accordingly.

Putting It All Together

Load the completed sketch for Project 9 from your Arduino Sketchbook and download it to the board (see Chapter 1).

Summary

In this chapter we have used a variety of LEDs and software techniques for lighting them in interesting ways. In Chapter 5 we will investigate some different types of sensors and use them to provide inputs to our projects.

CHAPTER 5

Sensor Projects

SENSORS TURN REAL-WORLD measurements into electronic signals that we can then use on our Arduino boards. The projects in this chapter are all about using light and temperature.

We also look at how to interface with keypads and rotary encoders.

Project 10
Keypad Security Code

This project would not be out of place in the lair of any Evil Genius worth their salt. A secret code must be entered on the keypad, and if it is correct, a green LED will light; otherwise, a red LED will stay lit. In Project 27 we will revisit this project and show how it cannot just show the appropriate light but also control a door lock.

COMPONENTS AND EQUIPMENT		
	Description	Appendix
	Arduino Uno or Leonardo	m1/m2
D1	5-mm red LED	s1
D2	5-mm green LED	s2
R1-2	270 Ω, 0.25 W resistor	r3
K1	4 by 3 keypad	h11
	0.1-inch header strip	h12
	Solderless breadboard	h1
	Jumper wires	h2

Unfortunately, keypads do not usually have pins attached, so we will have to attach some, and the only way to do that is to solder them on. So this is another of our projects where you will have to do a little soldering.

Hardware

The schematic diagram for Project 10 is shown in Figure 5-1. By now, you will be used to LEDs; the new component is the keypad.

Keypads are normally arranged in a grid so that when one of the keys is pressed, it connects a row to a column. Figure 5-2 shows a typical arrangement for a 12-key keypad with numbers from 0 to 9 and * and # keys.

The key switches are arranged at the intersection of row-and-column wires. When a key is pressed, it connects a particular row to a particular column.

By arranging the keys in a grid like this, it means that we only need to use 7 (4 rows + 3 columns) of our digital pins rather than 12 (one for each key).

However, it also means that we have to do a bit more work in the software to determine which keys are pressed. The basic approach we have to take is to connect each row to a digital output and each column to a digital input. We then put each output high in turn and see which inputs are high.

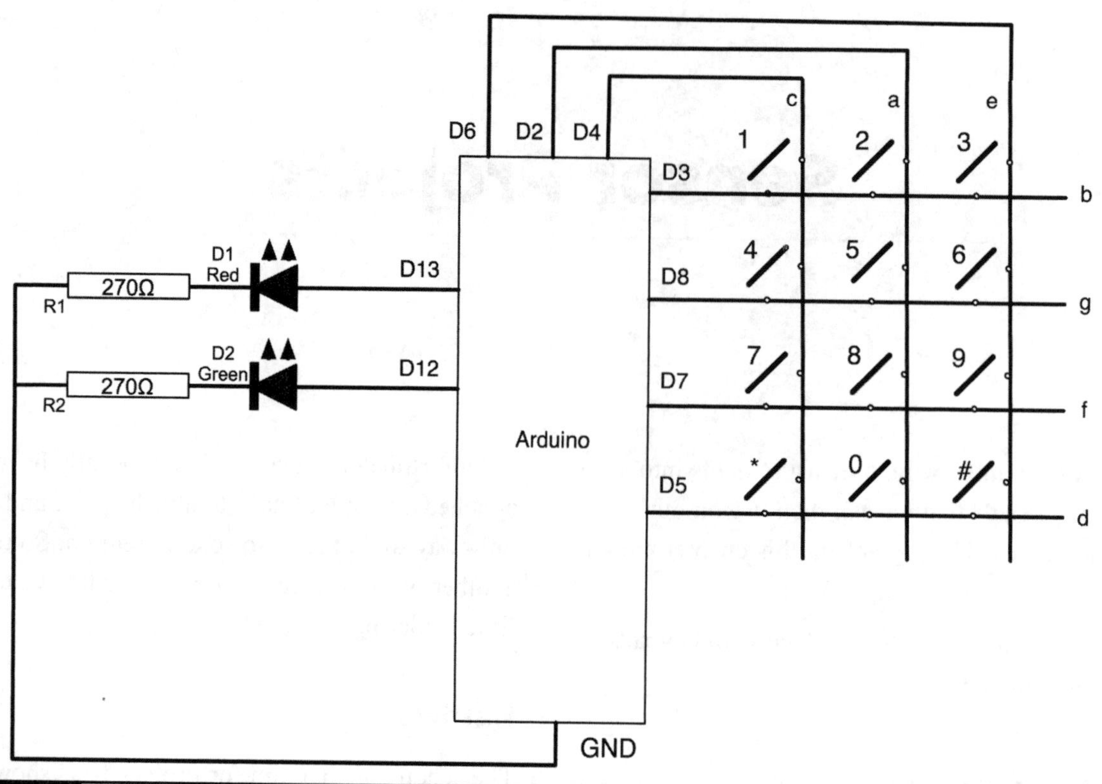

Figure 5-1 Schematic diagram for Project 10.

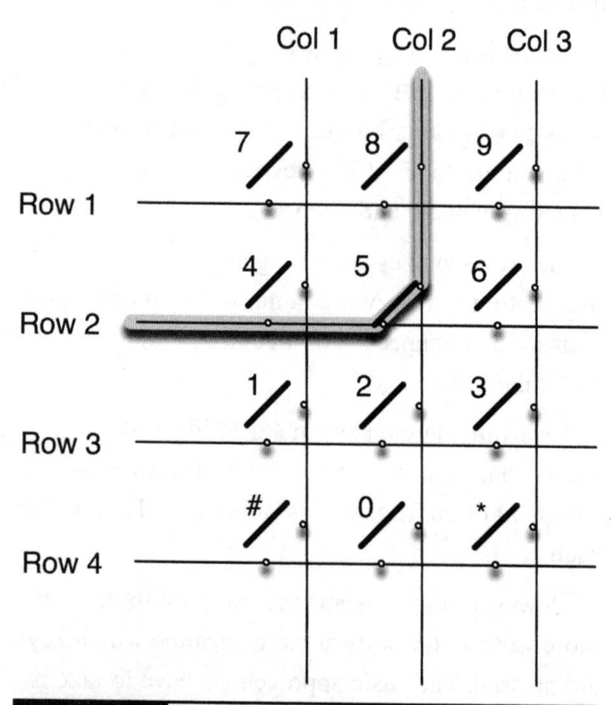

Figure 5-2 A 12-key keypad.

Figure 5-3 shows how you can solder seven pins from a pin header strip onto the keypad so that you can then connect it to the breadboard. Pin headers are bought in strips and can be easily snapped to provide the number of pins required.

Now we just need to find out which pin on the keypad corresponds to which row or column. If we are lucky, the keypad will come with a datasheet

Figure 5-3 Soldering pins to the keypad.

that tells us this. If not, we will have to do some detective work with a multimeter. Set the multimeter to continuity so that it beeps when you connect the leads together. Then get some paper, draw a diagram of the keypad connections, and label each pin with a letter from a to g. Then write a list of all the keys. Then, holding each key down in turn, find the pair of pins that make the multimeter beep, indicating a connection (Figure 5-4). Release the key to check that you have indeed found the correct pair. After a while, a pattern will emerge, and you will be able to see how the pins relate to rows and columns. Figure 5-4 shows the arrangement for the keypad used by the author.

The completed breadboard layout is shown in Figure 5-5 and the assembled breadboard in Figure 5-6. Note that your keypad may have a different pinout. If so, you will need to change the jumper wires connected to it accordingly.

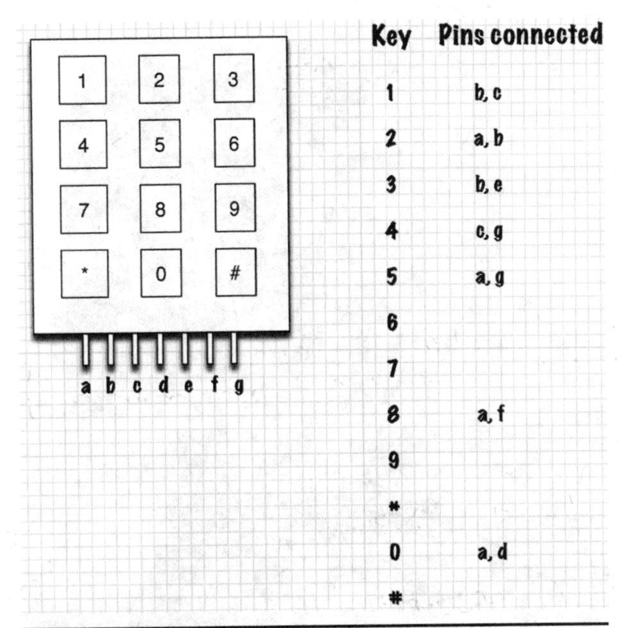

Figure 5-4 Working out the keypad connections.

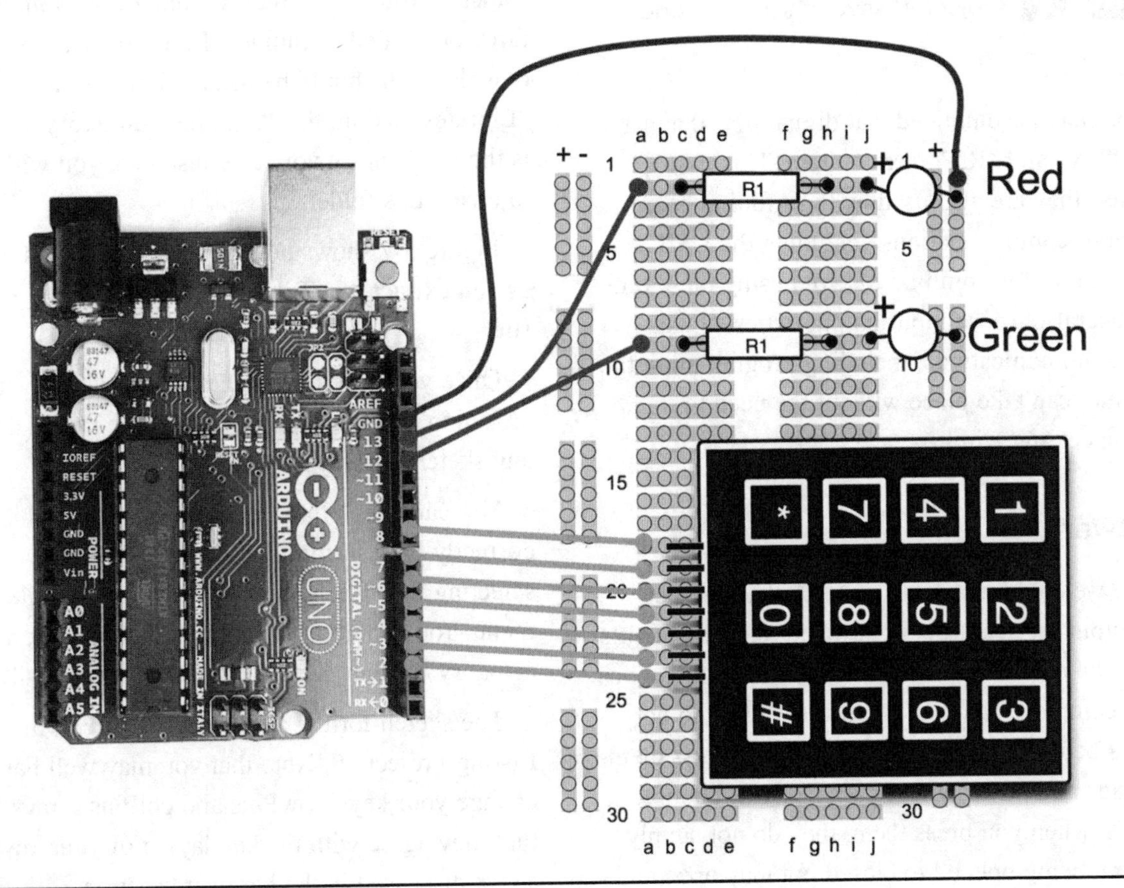

Figure 5-5 Project 10 breadboard layout.

Figure 5-6 Project 10 keypad security code.

You may have noticed that digital pins 0 and 1 have "TX" and "RX" next to them. This is so because they are also used by the Arduino board for serial communications, including the USB connection. It is common to avoid using these pins for general-purpose input-output duties so that serial communications, including programming the Arduino, can take place without the need to disconnect any wires.

Software

While we could just write a sketch that turns on the output for each row in turn and reads the inputs to get the coordinates of any key pressed, it is a bit more complex than that because switches do not always behave in a good way when you press them. Keypads and push switches are likely to bounce. That is, when you press them, they do not simply go from being opened to closed but may open and close several times as part of pressing the button.

Fortunately for us, Mark Stanley and Alexander Brevig have created a library that you can use to connect to keypads that handle such things for us. This is a good opportunity to demonstrate installing a library into the Arduino software.

In addition to the libraries that come with the Arduino, many people have developed their own libraries and published them for the benefit of the Arduino community. The Evil Genius is much amused by such altruism and sees it as a great weakness. However, the Evil Genius is not above using such libraries for his own devious ends.

To make use of this library, we must first download it from the Arduino website at this address: www.arduino.cc/playground/Code/Keypad.

Download the file Keypad.zip to your desktop.

Whether using Windows, Mac, or LINUX, you will find that the Arduino software has created a folder in your "Documents" folder that contains a directory called "Arduino." Libraries that you download all should be installed in a folder called "Libraies" within this "Arduino" directory. If this is the first library you have installed, you will need to create this folder.

Figure 5-7 shows how you can create this folder as you extract the "Library" folder from the Zip file.

Once you have installed this library into your "Arduino" directory, you will be able to use it with any sketches that you write.

You can check that the library is installed correctly by restarting the Arduino IDE and selecting the "Examples" option from the File menu. You should now find that there is a new category for the "Keypad" library (Figure 5-8).

The sketch for the application is shown in Listing Project 10. Note that you may well have to change your keys' rowPins and colPins arrays so that they agree with the key layout of your keypad, as we discussed in the hardware section.

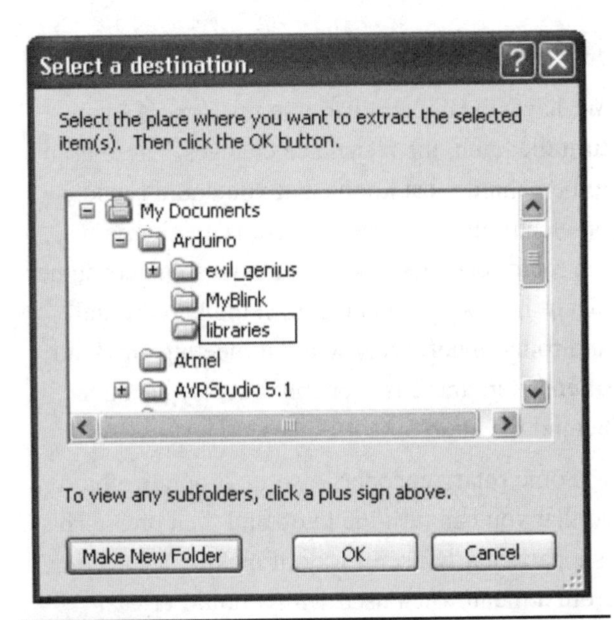

Figure 5-7 Unzipping the library for Windows.

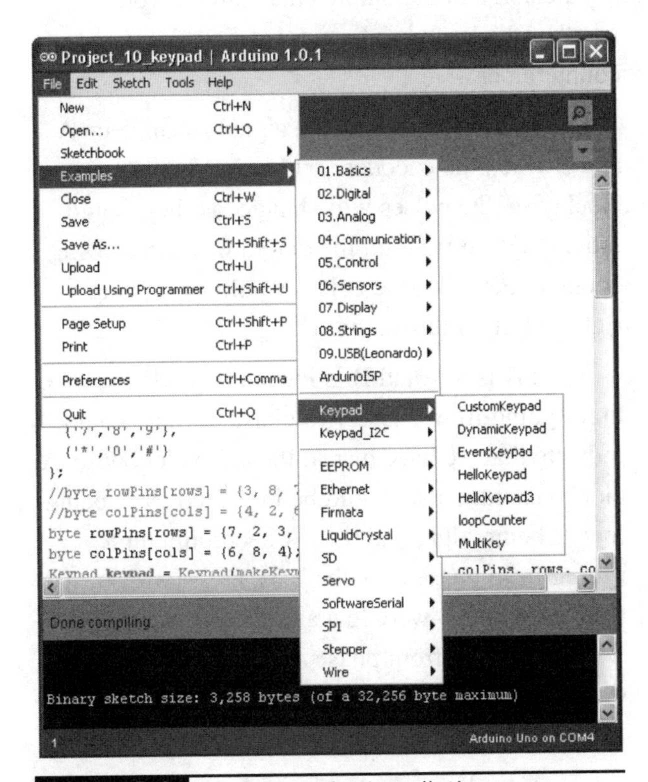

Figure 5-8 Checking the installation.

LISTING PROJECT 10

```
#include <Keypad.h>

char* secretCode = "1234";
int position = 0;

const byte rows = 4;
const byte cols = 3;
char keys[rows][cols] = {
  {'1','2','3'},
  {'4','5','6'},
  {'7','8','9'},
  {'*','0','#'}
};
byte rowPins[rows] = {7, 2, 3, 5};
byte colPins[cols] = {6, 8, 4};
Keypad keypad =
Keypad(makeKeymap(keys), rowPins,
       colPins, rows, cols);

int redPin = 13;
int greenPin = 12;

void setup()
{
  pinMode(redPin, OUTPUT);
  pinMode(greenPin, OUTPUT);
  setLocked(true);
}

void loop()
{
  char key = keypad.getKey();
  if (key == '*' || key == '#')
  {
    position = 0;
    setLocked(true);
  }
  if (key == secretCode[position])
  {
    position ++;
  }
  if (position == 4)
  {
    setLocked(false);
  }
```

(continued on next page)

LISTING PROJECT 10 (*continued*)

```
    delay(100);
}

void setLocked(int locked)
{
    if (locked)
    {
        digitalWrite(redPin, HIGH);
        digitalWrite(greenPin, LOW);
    }
    else
    {
        digitalWrite(redPin, LOW);
        digitalWrite(greenPin, HIGH);
    }
}
```

This sketch is quite straightforward. The loop function checks for a key press. If the key pressed is a # or a *, it sets the position variable back to 0. If, on the other hand, the key pressed is one of the numerals, it checks to see if the next key expected (secretCode[position]) is the key just pressed, and if it is, it increments position by one. Finally, the loop checks to see if position is 4, and if it is, it sets the LEDs to their unlocked state.

Putting It All Together

Load the completed sketch for Project 10 from your Arduino Sketchbook and download it to the board (see Chapter 1).

If you have trouble getting this to work, it is most likely a problem with the pin layout on your keypad. So persevere with the multimeter to map out the pin connections.

Rotary Encoders

We have already met variable resistors: As you turn the knob, the resistance changes. These used to be behind most knobs that you could twiddle on electronic equipment. There is an alternative, the rotary encoder, and if you own some consumer electronics where you can turn the knob round and round indefinitely without meeting any kind of end stop, there is probably a rotary encoder behind the knob.

Some rotary encoders also incorporate a button so that you can turn the knob and then press. This is a particularly useful way of making a selection from a menu when used with a liquid-crystal display (LCD) screen.

A rotary encoder is a digital device that has two outputs (A and B), and as you turn the knob, you get a change in the outputs that can tell you whether the knob has been turned clockwise or counterclockwise.

Figure 5-9 shows how the signals change on A and B when the encoder is turned. When rotating clockwise, the pulses will change, as they would moving left to right on the diagram; when moving counterclockwise, the pulses would be moving right to left on the diagram.

So, if A is low and B is low and then B becomes high (going from phase 1 to phase 2), that would indicate that we have turned the knob clockwise. A clockwise turn also would be indicated by A being low, B being high, and then A becoming high (going from phase 2 to phase 3), etc. However, if A were high and B were low and then B went high, we have moved from phase 4 to phase 3 and are therefore turning counterclockwise.

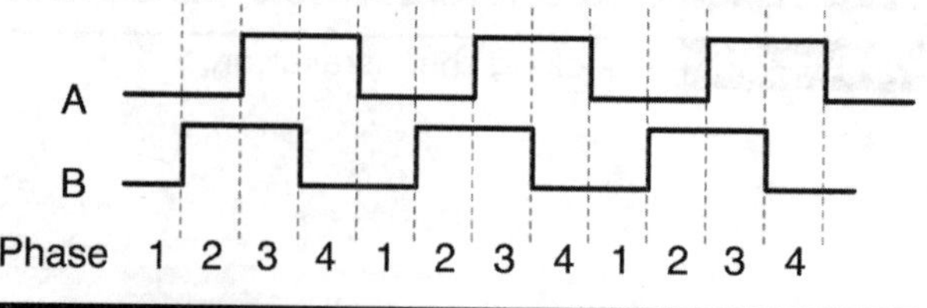

Figure 5-9 Pulses from a rotary encoder.

Project 11
Model Traffic Signal Using a Rotary Encoder

This project uses a rotary encoder with a built-in push switch to control the sequence of the traffic signals and is based on Project 5. It is a much more realistic version of a traffic signal controller and is really not far off the logic that you would find in a real traffic signal controller.

Rotating the rotary encoder will change the frequency of the light sequencing. Pressing the button will test the lights, turning them all on at the same time, while the button is pressed.

The components are the same as for Project 5, with the addition of the rotary encoder in place of the original push switch.

COMPONENTS AND EQUIPMENT

	Description	Appendix
	Arduino Uno or Leonardo	m1/m2
D1	5-mm red LED	s1
D2	5-mm yellow LED	s3
D3	5-mm green LED	s2
R1-R3	270 Ω, 0.25 W resistor	r3
S1	Rotary encoder with push switch	h13
	Solderless breadboard	h1
	Jumper wires	h2

Hardware

The schematic diagram for Project 11 is shown in Figure 5-10. The majority of the circuitry is the same as for Project 5, except that now we have a rotary encoder.

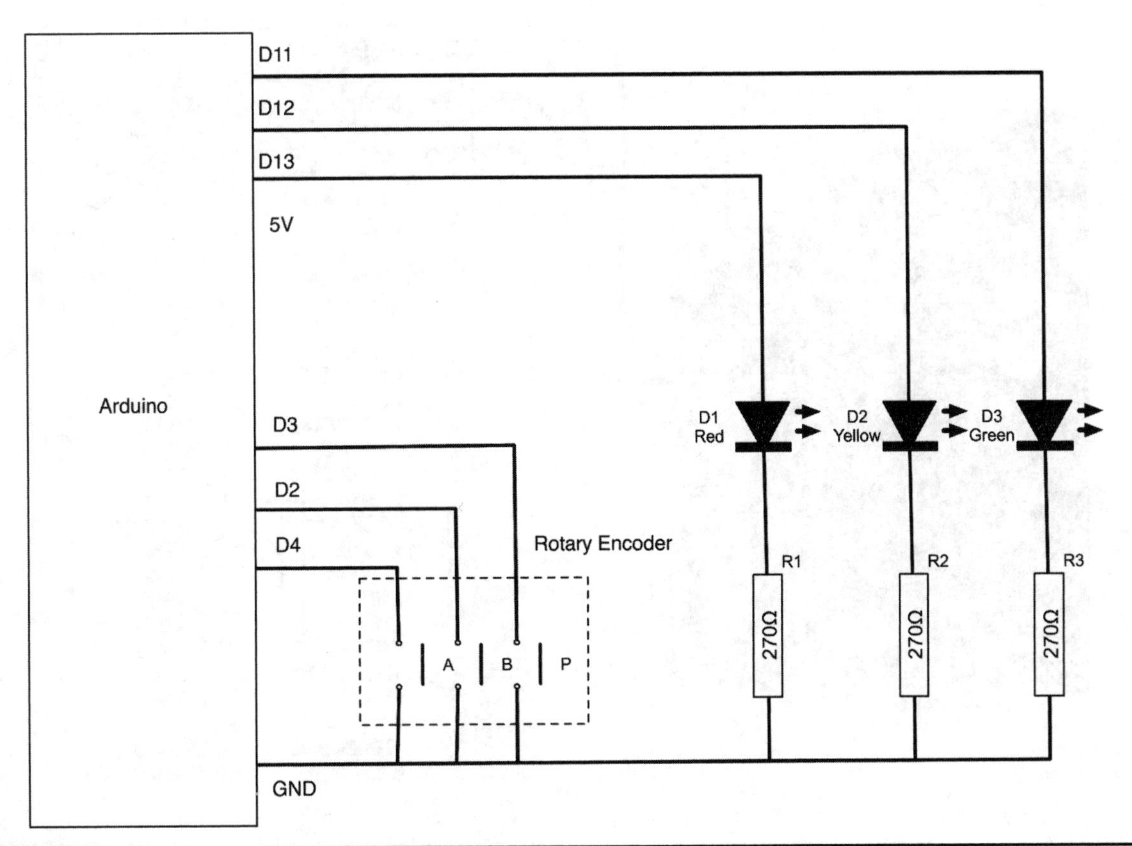

Figure 5-10 Schematic diagram for Project 11.

The rotary encoder works just as if there were three switches: one each for A and B and one for the push switch.

Since the schematic is much the same as for Project 5, it will not be much of a surprise to see that the breadboard layout (Figure 5-11) is also similar to the one for that project.

Software

The starting point for the sketch is the sketch for Project 5. We have added code to read the encoder and to respond to the button press by turning all the LEDs on. We also have taken the opportunity to enhance the logic behind the lights to make them behave in a more realistic way, changing automatically. In Project 5, when you held down the button, the lights changed sequence roughly once per second. In a real traffic signal, the lights stay green and red a lot longer than they are yellow. So our sketch now has two periods: shortPeriod, which does not alter but is used when the lights are changing, and longPeriod, which determines how long they are illuminated when green or red. This longPeriod is the period that is changed by turning the rotary encoder.

The key to handling the rotary encoder lies in the function getEncoderTurn. Every time this is called, it compares the previous state of A and B with their current state, and if something has changed, it works out whether it was clockwise or counterclockwise and returns a –1 or 1, respectively. If there is no change (the knob has not been turned), it returns 0. This function must be called frequently, or turning the rotary controller quickly will result in some changes not being recognized correctly.

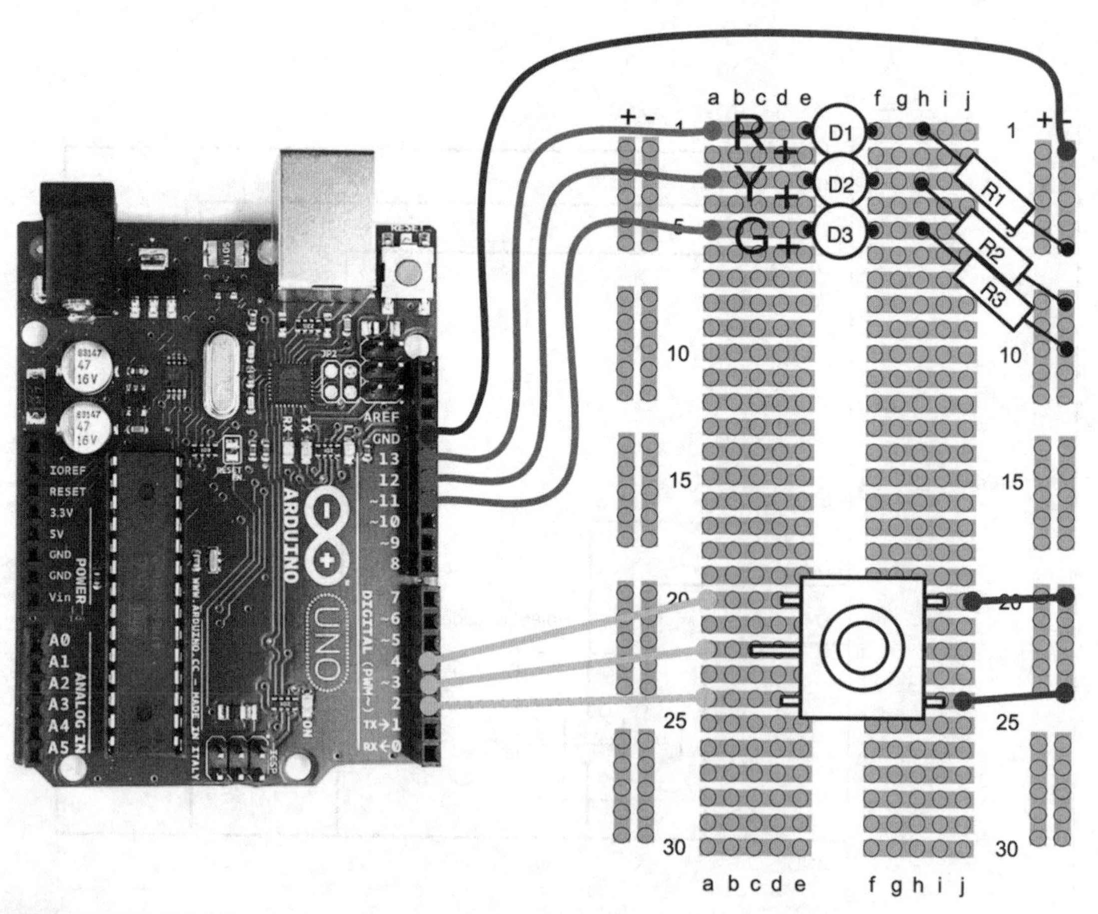

Figure 5-11 Breadboard layout for Project 11.

If you want to use a rotary encoder for other projects, you can just copy this function. The function uses the static modifier for the oldA and oldB variables. This is a useful technique that allows the function to retain the values between one call of the function and the next, where normally it would reset the value of the variable every time the function is called.

LISTING PROJECT 11

```
int redPin = 13;
int yellowPin = 12;
int greenPin = 11;
int aPin = 4;
int bPin = 2;
int buttonPin = 3;

int state = 0;
int longPeriod = 5000;          // Time at green or red
int shortPeriod = 700;          // Time period when changing
int targetCount = shortPeriod;
int count = 0;

void setup()
{
  pinMode(aPin, INPUT_PULLUP);
  pinMode(bPin, INPUT_PULLUP);
  pinMode(buttonPin, INPUT_PULLUP);
  pinMode(redPin, OUTPUT);
  pinMode(yellowPin, OUTPUT);
  pinMode(greenPin, OUTPUT);
}

void loop()
{
  count++;
  if (digitalRead(buttonPin) == LOW)
  {
    setLights(HIGH, HIGH, HIGH);
  }
  else
  {
    int change = getEncoderTurn();
    int newPeriod = longPeriod + (change * 1000);
    if (newPeriod >= 1000 && newPeriod <= 10000)
    {
      longPeriod = newPeriod;
    }
    if (count > targetCount)
    {
```

(continued on next page)

LISTING PROJECT 11 (continued)

```
        setState();
        count = 0;
      }
    }
    delay(1);
}

int getEncoderTurn()
{
    // return -1, 0, or +1
    static int oldA = LOW;
    static int oldB = LOW;
    int result = 0;
    int newA = digitalRead(aPin);
    int newB = digitalRead(bPin);
    if (newA != oldA || newB != oldB)
    {
      // something has changed
      if (oldA == LOW && newA == HIGH)
      {
        result = -(oldB * 2 - 1);
      }
    }
    oldA = newA;
    oldB = newB;
    return result;
}

int setState()
  {
    if (state == 0)
    {
      setLights(HIGH, LOW, LOW);
      targetCount = longPeriod;
      state = 1;
    }
    else if (state == 1)
    {
      setLights(HIGH, HIGH, LOW);
      targetCount = shortPeriod;
      state = 2;
    }
    else if (state == 2)
    {
      setLights(LOW, LOW, HIGH);
      targetCount = longPeriod;
      state = 3;
```

LISTING PROJECT 11 *(continued)*

```
    }
    else if (state == 3)
    {
      setLights(LOW, HIGH, LOW);
      targetCount = shortPeriod;
      state = 0;
     }
  }

void setLights(int red, int yellow, int green)
{
  digitalWrite(redPin, red);
  digitalWrite(yellowPin, yellow);
  digitalWrite(greenPin, green);
}
```

This sketch illustrates a useful technique that lets you time events (turning an LED on for so many seconds) while at the same time checking the rotary encoder and button to see if they have been turned or pressed. If we just used the Arduino delay function with, say, 20,000, for 20 seconds, we would not be able to check the rotary encoder or switch in that period.

So what we do is use a very short delay (1 millisecond) but maintain a count that is incremented each time round the loop. Thus, if we want to delay for 20 seconds, we stop when the count has reached 20,000. This is less accurate than a single call to the delay function because the 1 millisecond is actually 1 millisecond plus the processing time for the other things that are done inside the loop.

Putting It All Together

Load the completed sketch for Project 11 from your Arduino Sketchbook and download it to the board (see Chapter 1).

You can press the rotary encoder button to test the LEDs and turn the rotary encoder to change how long the signal stays green and red.

Sensing Light

A common and easy-to-use device for measuring light intensity is the light-dependent resistor (or LDR). They are also sometimes called *photoresistors*.

The brighter the light falling on the surface of the LDR, the lower is the resistance. A typical LDR will have a dark resistance of up to 2 MΩ and a resistance when illuminated in bright daylight of perhaps 20 kΩ.

We can convert this change in resistance to a change in voltage by using the LDR, with a fixed resistor as a voltage divider, connected to one of our analog inputs. The schematic for this is shown in Figure 5-12.

With a fixed resistor of 100K, we can do some rough calculations about the voltage range to expect at the analog input.

In darkness, the LDR will have a resistance of 2 MΩ, so with a fixed resistor of 100K, there will be about a 20:1 ratio of voltage, with most of that voltage across the LDR, so this would mean about 4V across the LDR and 1V at the analog pin.

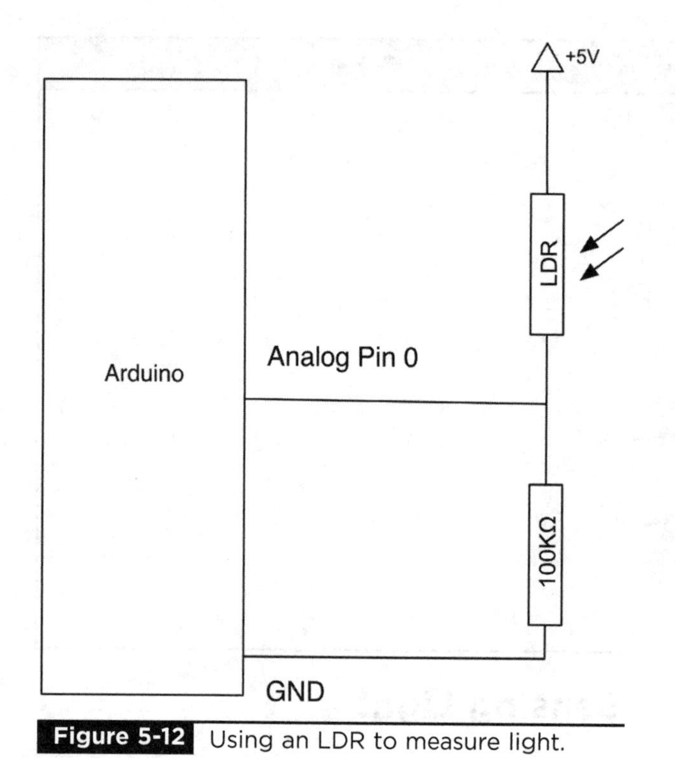

Figure 5-12 Using an LDR to measure light.

On the other hand, if the LDR is in bright light, its resistance might fall to 20 kΩ. The ratio of voltages then would be about 4:1 in favor of the fixed resistor, giving a voltage at the analog input of about 4V.

A more sensitive photo detector is the phototransistor. This functions like an ordinary transistor except there is not usually a base connection. Instead, the collector current is controlled by the amount of light falling on the phototransistor.

Project 12
Pulse-Rate Monitor

This project uses an ultrabright infrared (IR) LED and a phototransistor to detect the pulse in your finger. It then flashes a red LED in time with your pulse.

COMPONENTS AND EQUIPMENT		
	Description	Appendix
	Arduino Uno or Leonardo	m1/m2
D1	5-mm red LED	s1
D2	5-mm IR LED sender 940 nm	s20
R1	56 kΩ, 0.25 W resistor	r7
R2	270 Ω, 0.25 W resistor	r3
R4	100 Ω, 0.25 W resistor	r2
T1	IR phototransistor (same wavelength as D2)	s19
	Solderless breadboard	h1
	Jumper wires	h2

Hardware

The pulse monitor works as follows: Shine the bright LED onto one side of your finger while the phototransistor on the other side of your finger picks up the amount of transmitted light. The resistance of the phototransistor will vary slightly as the blood pulses through your finger.

The schematic for this is shown in Figure 5-13 and the breadboard layout in Figure 5-15. We have chosen quite a high value of resistance for R1 because most of the light passing through the finger will be absorbed, and we want the phototransistor to be quite sensitive. You may need to experiment with the value of the resistor to get the best results.

It is important to shield the phototransistor from as much stray light as possible. This is particularly important for domestic lights that actually fluctuate at 50 or 60 Hz and will add a considerable amount of noise to our weak heart signal.

For this reason, the phototransistor and LED are built into a tube or corrugated cardboard held together with duct tape. The construction of this is shown in Figure 5-14.

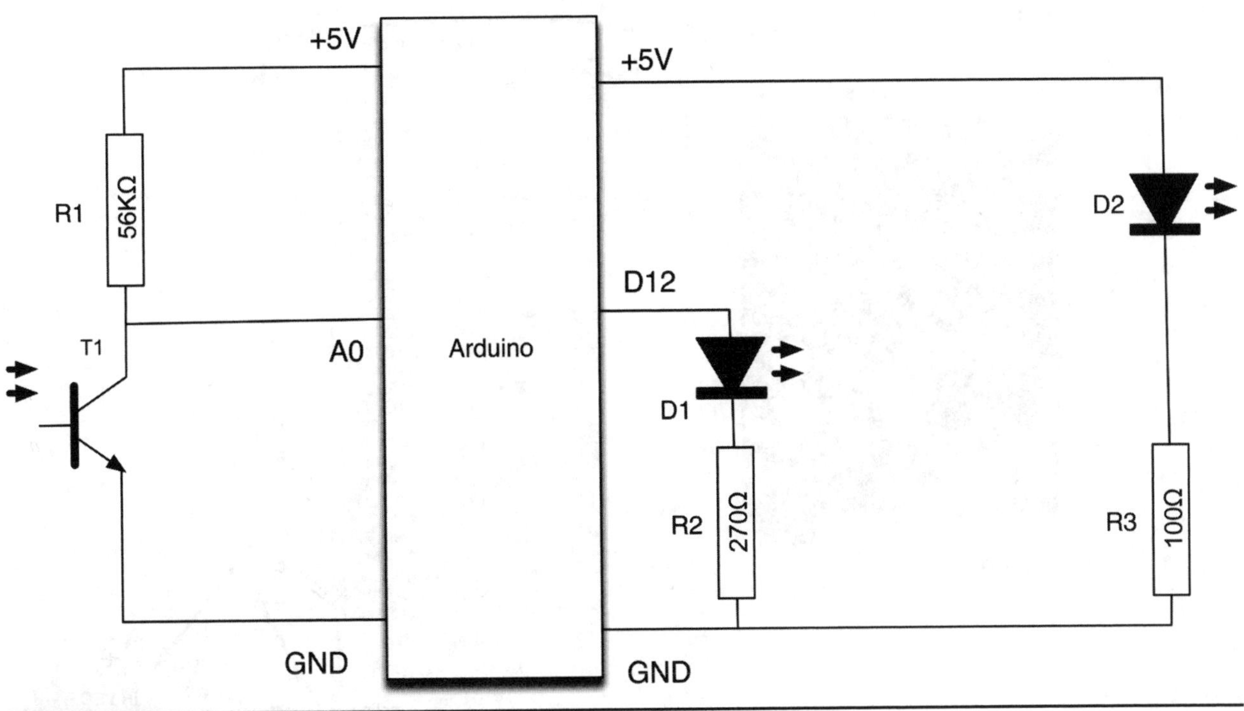

Figure 5-13 Schematic for Project 12.

Two 5-mm holes are drilled opposite each other in the tube, and the LED is inserted into one side and the phototransistor into the other. Short leads are soldered to the LED and phototransistor, and then another layer of tape is wrapped over everything to hold it all in place. Be sure to check which colored wire is connected to which lead of the LED and phototransistor before you tape them up.

Figure 5-14 Sensor tube for heart monitor.

It is also a good idea to use screened wire for the phototransistor to reduce interference. It is also worth noting that a peculiarity of most IR LEDs is that the longer lead is negative rather than positive, so check the data sheet of the device before you attach it.

The breadboard layout for this project (Figure 5-15) is very straightforward.

The final "finger tube" can be seen in Figure 5-16.

Software

The software for this project is quite tricky to get right. Indeed, the first step is not to run the entire final script but rather a test script that will gather data that we can then paste into a spreadsheet and chart to test out the smoothing algorithm (more on this later).

The test script is provided in Listing Project 12.

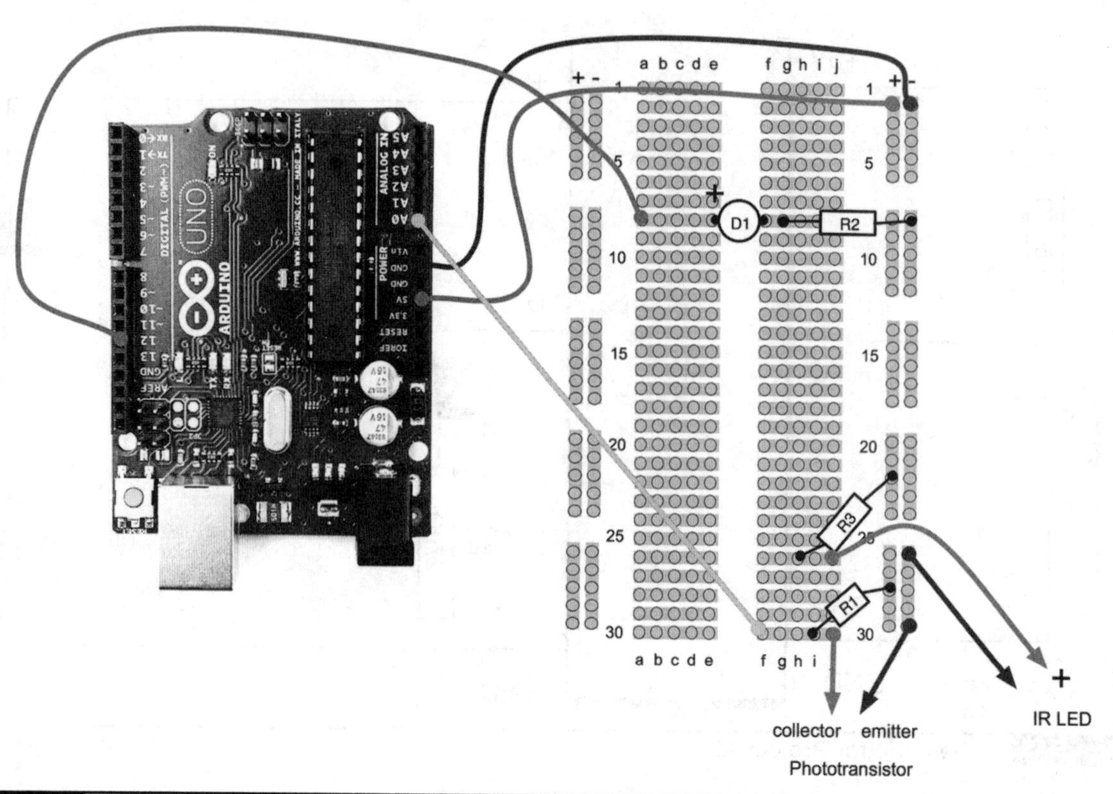

Figure 5-15 Breadboard layout for Project 12.

Figure 5-16 Project 12: pulse-rate monitor.

LISTING PROJECT 12—TEST SCRIPT

```
int ledPin = 13;
int sensorPin = A0;

double alpha = 0.75;
int period = 20;
double change = 0.0;

void setup()
{
  pinMode(ledPin, OUTPUT);
  Serial.begin(115200);
}

void loop()
{
    static double oldValue = 0;
    static double oldChange = 0;
    int rawValue =
analogRead(sensorPin);
    double value = alpha * oldValue
    + (1 - alpha) * rawValue;

    Serial.print(rawValue);
    Serial.print(",");
    Serial.println(value);

    oldValue = value;
    delay(period);
}
```

This script reads the raw signal from the analog input, applies the smoothing function, and then writes both values to the Serial Monitor, where we can capture them and paste them into a spreadsheet for analysis. Note that the Serial Monitor's communications is set to its fastest rate to minimize the effects of the delays caused by sending the data. When you start the Serial Monitor, you will need to change the serial speed to 115,200 baud.

The smoothing function uses a technique called *leaky integration*, and you can see in the code where we do this smoothing using the line

```
double value = alpha * oldValue +
(1 - alpha) * rawValue;
```

The variable alpha is a number greater than 0 but less than 1 and determines how much smoothing to do.

Put your finger into the sensor tube, start the Serial Monitor, and leave it running for 3 or 4 seconds to capture a few pulses.

Then copy and paste the captured text into a spreadsheet. You will probably be asked for the column delimiter character, which is a comma. The resulting data and a line chart drawn from the two columns are shown in Figure 5-17.

The more jagged trace is from the raw data read from the analog port, and the smoother trace clearly has most of the noise removed. If the smoothed trace shows significant noise—in particular, any false peaks that will confuse the monitor—increase the level of smoothing by decreasing the value of alpha.

Once you have found the right value of alpha for your sensor arrangement, you can transfer this value into the real sketch and switch over to using the real sketch rather than the test sketch. The real sketch is provided in the following listing.

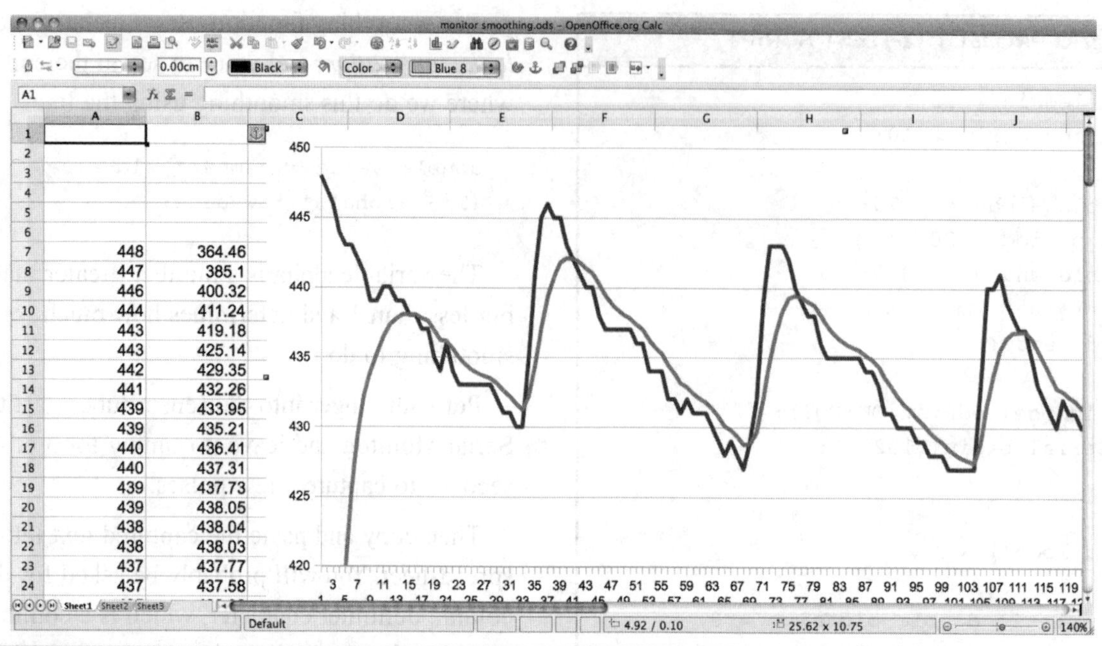

Figure 5-17 Heart monitor test data pasted into a spreadsheet.

LISTING PROJECT 12

```
int ledPin = 12;
int sensorPin = 0;

double alpha = 0.75;
int period = 20;
double change = 0.0;

void setup()
{
  pinMode(ledPin, OUTPUT);
}

void loop()
{
    static double oldValue = 0;
    static double oldChange = 0;
    int rawValue =
analogRead(sensorPin);
    double value = alpha * oldValue
      + (1 - alpha) * rawValue;
    change = value - oldValue;

    digitalWrite(ledPin, (change <
      0.0 && oldChange > 0.0));

    oldValue = value;
    oldChange = change;
    delay(period);
}
```

There now just remains the problem of detecting the peaks. Looking at Figure 5-17, we can see that if we keep track of the previous reading, we can see that the readings are gradually increasing until the change in reading flips over and becomes negative. So, if we lit the LED whenever the old change was positive but the new change was negative, we would get a brief pulse from the LED at the peak of each pulse.

Putting It All Together

Both the test and real sketch for Project 12 are in your Arduino Sketchbook. For instructions on downloading it to the board, see Chapter 1.

As mentioned earlier, getting this project to work is a little tricky. You will probably find that you have to get your finger in just the right place to start getting a pulse. If you are having trouble, run the test script as described previously to check that your detector is getting a pulse and the smoothing factor alpha is low enough.

The author would like to point out that this device should not be used for any kind of real medical application.

Measuring Temperature

Measuring temperature is a similar problem to measuring light intensity. Instead of an LDR, a device called a *thermistor* is used. As the temperature increases, so does the resistance of the thermistor.

When you buy a thermistor, it will have a stated resistance. In this case, the thermistor chosen is 33 kΩ. This will be the resistance of the device at 25°C.

The formula for calculating the resistance at a particular temperature is given by

$$R = R_o \exp(-beta/(T + 273) - beta/(T_o + 273)$$

You can do the math if you like, but a much simpler way to measure temperature is to use a special-purpose thermometer chip such as the TMP36. This three-pinned device has two pins for the power supply (5V) and a third output pin, whose temperature T in degrees C is related to the output voltage V by the equation

$$T = (V - 0.5) \times 100$$

So, if the voltage at its output is 1V, the temperature is 50°C.

Project 13
USB Temperature Logger

This project is controlled by your computer, but once given its logging instructions, the device can be disconnected and run on batteries to do its logging. While logging, it stores its data, and then when the logger is reconnected, it will transfer its data back over the USB connection, where it can be imported into a spreadsheet. By default, the logger will record one sample every 5 minutes and can record up to 1000 samples.

To instruct the temperature logger from your computer, we have to define some commands that can be issued from the computer. These are shown in Table 5-1.

TABLE 5-1 Temperature Logger Commands

R	Read the data out of the logger as CSV text
X	Clear all data from the logger
C	Centigrade mode
F	Fahrenheit mode
1-9	Set the sample period in minutes from 1 to 9
G	Go! Start logging temperatures
?	Reports the status of the device, number of samples taken, etc.

This project just requires a TMP36 that can fit directly into the sockets on the Arduino.

COMPONENTS AND EQUIPMENT

	Description	Appendix
	Arduino Uno or Leonardo	m1/m2
IC1	TMP36	s22

Hardware

The schematic diagram for Project 13 is shown in Figure 5-18.

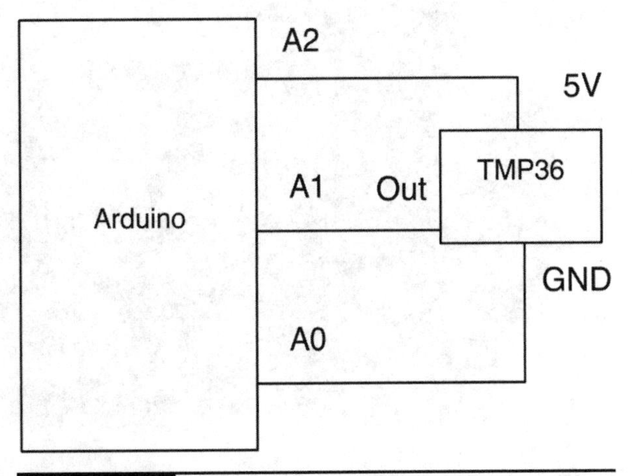

Figure 5-18 Schematic diagram for Project 13.

This is so simple that we can simply fit the leads of the TMP36 into the Arduino board, as shown in Figure 5-19. Note that the curved side of the TMP36 should face outward from the Arduino. Putting a little kink in the leads with pliers will ensure a better contact.

Two of the analog pins (A0 and A2) are going to be used for the GND and 5V power connections to the TMP36. The TMP36 uses very little current, so the pins can easily supply enough to power it if we set one pin HIGH and the other LOW.

Software

The software for this project is a little more complex than for some of our other projects (see Listing Project 13). All the variables that we have used in our sketches so far are forgotten as soon as the Arduino board is reset or disconnected from the power. Sometimes we want to be able to store data persistently so that it is there next time we start up the board. This can be done by using the special type of memory on the Arduino called *EEPROM*, which stands for *electrically erasable programmable read-only memory*. The Arduino Uno and Leonardo both have 1024 bytes of EEPROM.

For the data logger to be useful, it needs to remember the readings that it has already taken, even when it is disconnected from the computer and powered from a battery. It also needs to remember the logging period.

This is the first project where we have used the Arduino's EEPROM to store values so that they are not lost if the board is reset or disconnected from the power. This means that once we have set our data-logging recording, we can disconnect it from the USB lead and leave it running on batteries. Even if the batteries go dead, our data will still be there the next time we connect it.

Figure 5-19 Project 13: temperature logger.

LISTING PROJECT 13

```
// Project 13 - Temperature Logger
#include <EEPROM.h>

#define analogPin 1
#define gndPin A0
#define plusPin A2
#define maxReadings 1000

int lastReading = 0;

boolean loggingOn;
//long period = 300;
long period = 10000; // 10 seconds
long lastLoggingTime = 0;
char mode = 'C';

void setup()
{
  pinMode(gndPin, OUTPUT);
  pinMode(plusPin, OUTPUT);
  digitalWrite(gndPin, LOW);
  digitalWrite(plusPin, HIGH);

  Serial.begin(9600);
  Serial.println("Ready");

  lastReading = EEPROM.read(0);              // First byte is reading position
  char sampleCh = (char)EEPROM.read(1);      // Second is logging period '0' to '9'
  if (sampleCh > '0' && sampleCh <= '9')
  {
    setPeriod(sampleCh);
  }
  loggingOn = true;                          // start logging on turn on
}

void loop()
{
  if (Serial.available())
  {
    char ch = Serial.read();
    if (ch == 'r' || ch == 'R')
    {
      sendBackdata();
    }
    else if (ch == 'x' || ch == 'X')
    {
```

(continued on next page)

LISTING PROJECT 13 (*continued*)

```
      lastReading = 0;
      EEPROM.write(0, 0);
      Serial.println("Data cleared");
    }
    else if (ch == 'g' || ch == 'G')
    {
      loggingOn = true;
      Serial.println("Logging started");
    }
    else if (ch > '0' && ch <= '9')
    {
      setPeriod(ch);
    }
    else if (ch == 'c' or ch == 'C')
    {
      Serial.println("Mode set to deg C");
      mode = 'C';
    }
    else if (ch == 'f' or ch == 'F')
    {
      Serial.println("Mode set to deg F");
      mode = 'F';
    }
    else if (ch == '?')
    {
      reportStatus();
    }
  }
  long now = millis();
  if (loggingOn && (now > lastLoggingTime + period))
  {
    logReading();
    lastLoggingTime = now;
  }
}

void sendBackdata()
{
  loggingOn = false;
  Serial.println("Logging stopped");
  Serial.println("------ cut here ------");
  Serial.print("Time (min)\tTemp (");
  Serial.print(mode);
  Serial.println(")");
  for (int i = 0; i < lastReading + 2; i++)
  {
      Serial.print((period * i) / 60000);
```

LISTING PROJECT 13 (*continued*)

```
      Serial.print("\t");
      float temp = getReading(i);
      if (mode == 'F')
      {
        temp = (temp * 9) / 5 + 32;
      }
      Serial.println(temp);
  }
  Serial.println("------ cut here ------");
}

void setPeriod(char ch)
{
  EEPROM.write(1, ch);
  long periodMins = ch - '0';
  Serial.print("Sample period set to: ");
  Serial.print(periodMins);
  Serial.println(" mins");
  period = periodMins * 60000;
}

void logReading()
{
  if (lastReading < maxReadings)
  {
    storeReading(measureTemp(), lastReading);
    lastReading++;
  }
  else
  {
    Serial.println("Full! logging stopped");
    loggingOn = false;
  }
}

float measureTemp()
{
  int a = analogRead(analogPin);
  float volts = a / 205.0;
  float temp = (volts - 0.5) * 100;
  return temp;
}

void storeReading(float reading, int index)
{
  EEPROM.write(0, (byte)index); // store the number of samples in byte 0
```

(continued on next page)

LISTING PROJECT 13 (*continued*)

```
  byte compressedReading = (byte)((reading + 20.0) * 4);
  EEPROM.write(index + 2, compressedReading);
  reportStatus();
}

float getReading(int index)
{
  lastReading = EEPROM.read(0);
  byte compressedReading = EEPROM.read(index + 2);
  float uncompressesReading = (compressedReading / 4.0) - 20.0;
  return uncompressesReading;
}

void reportStatus()
{
  Serial.println("----------------");
  Serial.println("Status");
  Serial.print("Current Temp C");
  Serial.println(measureTemp());
  Serial.print("Sample period (s)\t");
  Serial.println(period / 1000);
  Serial.print("Num readings\t");
  Serial.println(lastReading);
  Serial.print("Mode degrees\t");
  Serial.println(mode);
  Serial.println("----------------");
}
```

You will notice that at the top of this sketch we use the command #define for what in the past we would have used variables for. This is actually a more efficient way of defining constants—that is, values that will not change during the running of the sketch. So it is actually ideal for pin settings and constants such as beta. The command #define is what is called a *preprocessor directive*, and what happens is that just before the sketch is compiled, all occurrences of its name anywhere in the sketch are replaced by its value. It is very much a matter of personal taste whether you use #define or a variable.

Fortunately, reading and writing EEPROM happens just 1 byte at a time. So, if we want to write a variable that is a byte or a char, we can just use the functions EEPROM.write and EEPROM.read, as shown in the example here:

```
char letterToWrite = 'A';
EEPROM.write(0, myLetter);

char letterToRead;
letterToRead = EEPROM.read(0);
```

The 0 in the parameters for read and write is the address in the EEPROM to use. This can be any number between 0 and 1023, with each address being a location where 1 byte is stored.

In this project we want to store both the position of the last reading taken (in the lastReading variable) and all the readings. So we will record lastReading in the first byte of EEPROM, the logging period as a character 1 to 9, and then the actual reading data in the bytes that follow.

Each temperature reading is kept in a float, and if you remember from Chapter 2, a float occupies 4 bytes of data. Here we had a choice: We could either store all 4 bytes or find a way to encode the temperature into a single byte. We decided to take the latter route so that we can store as many readings as possible in the EEPROM.

The way we encode the temperature into a single byte is to make some assumptions about our temperatures. First, we assume that any temperature in Centigrade will be between –20 and +40. Anything higher or lower would likely damage our Arduino board anyway. Second, we assume that we only need to know the temperature to the nearest quarter of a degree.

With these two assumptions, we can take any temperature value we get from the analog input, add 20 to it, multiply it by 4, and still be sure that we always have a number between 0 and 240. Since a byte can hold a number between 0 and 255, that just fits nicely.

When we take our numbers out of EEPROM, we need to convert them back to a float, which we can do by reversing the process, dividing by 4, and then subtracting 20.

Both encoding and decoding the values are wrapped up in the functions storeReading and getReading. So, if we decided to take a different approach to storing the data, we would only have to change these two functions.

Putting It All Together

Load the completed sketch for Project 13 from your Arduino Sketchbook and download it to the board (see Chapter 1).

Now open the Serial Monitor (Figure 5-20), and for test purposes, we will set the temperature logger to log every minute by typing 1 in the Serial Monitor. The board should respond with the message "Sample period set to: 1 min." If we wanted to, we could then change the mode to Fahrenheit by typing **F** into the Serial Monitor. Now we can check the status of the logger by typing **?** (Figure 5-21).

In order to unplug the USB cable, we need to have an alternative source of power, such as the battery lead we made back in Project 6. You need

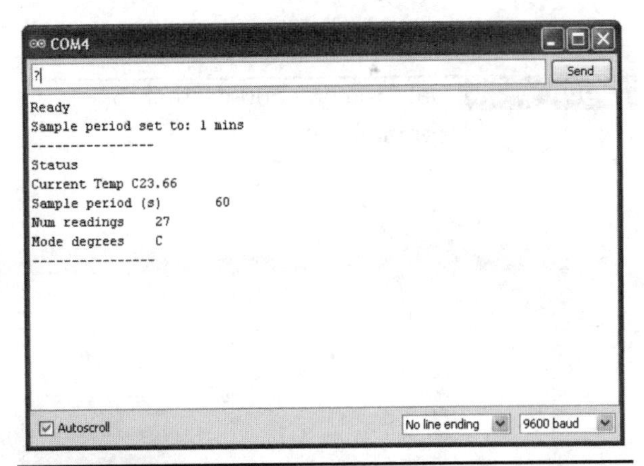

Figure 5-20 Issuing commands through the Serial Monitor.

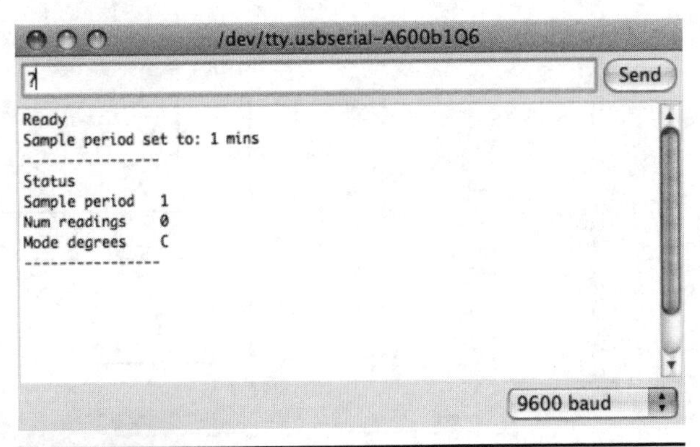

Figure 5-21 Displaying the Temperature Logger Status.

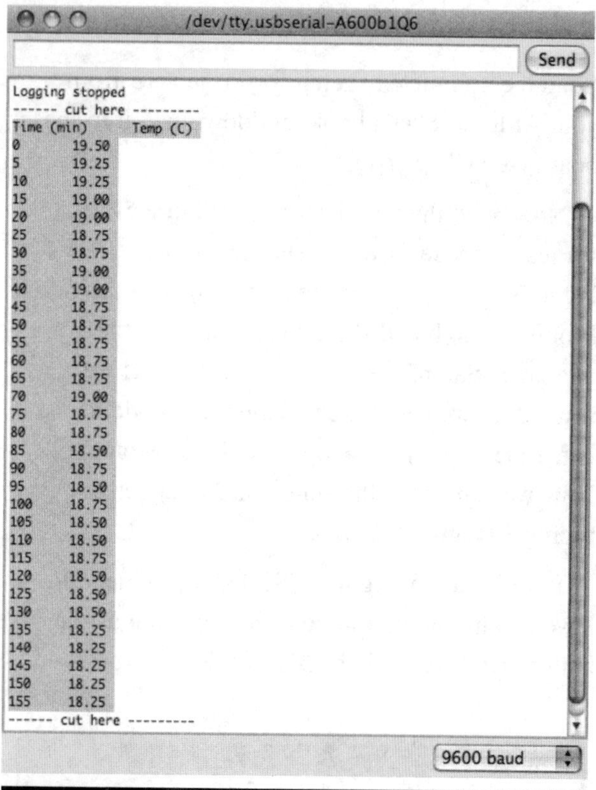

Figure 5-22 Data to copy and paste into a spreadsheet.

to have this plugged in and powered up at the same time as the USB connector is connected if you want the logger to keep logging after you disconnect the USB lead.

Finally, we can type the G command to start logging. We can then unplug the USB lead and leave our logger running on batteries. After waiting 10 or 15 minutes, we can plug it back in to see what data we have by opening the Serial Monitor and typing the R command, the results of which are shown in Figure 5-22. Select all the data, including the "Time" and "Temp" headings at the top.

Copy the text to the clipboard (press CTRL-C on Windows and LINUX, ALT-C on Macs), open a spreadsheet in a program such as Microsoft Excel, and paste it into a new spreadsheet (Figure 5-23).

Once in the spreadsheet, we can even draw a chart using our data.

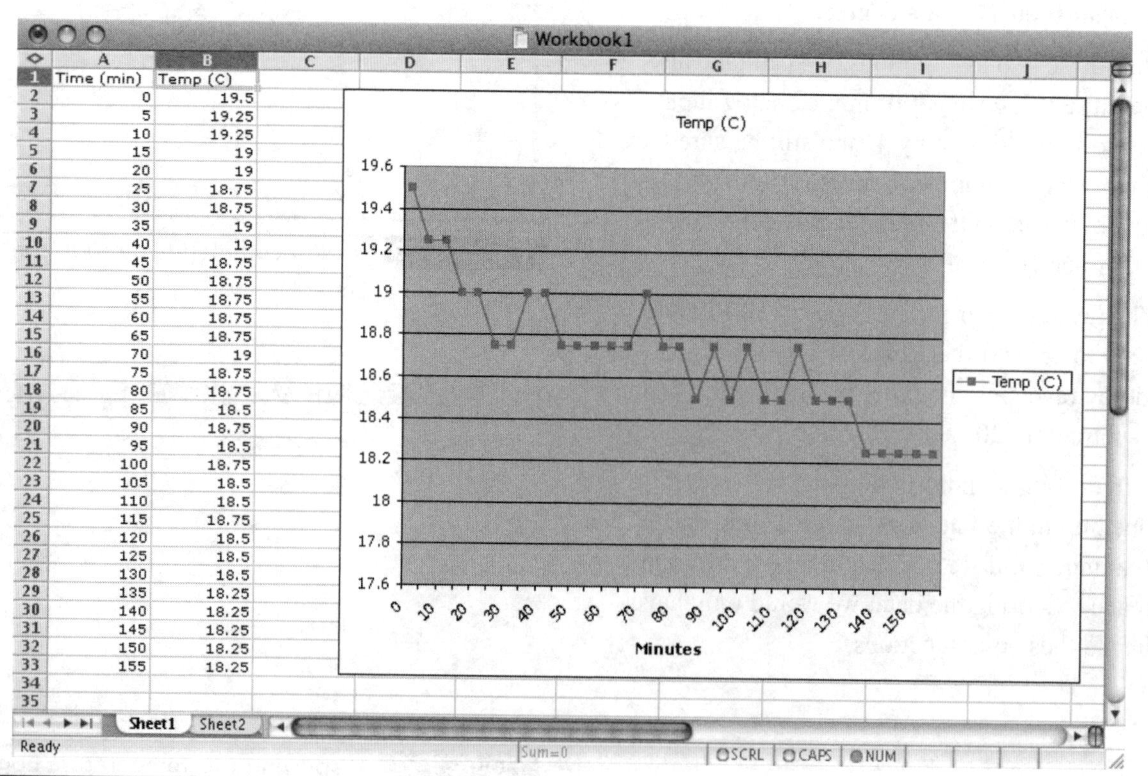

Figure 5-23 Temperature data imported into a spreadsheet.

Summary

We now know how to handle various types of sensors and input devices to go with our knowledge of LEDs. In the next section we will look at a number of projects that use light in various ways and get our hands on some more advanced display technologies, such as LCD text panels and seven-segment LEDs.

CHAPTER 6

Light Projects

IN THIS CHAPTER WE LOOK AT some more projects based on lights and displays. In particular, we look at how to use multicolor LEDs, seven-segment LEDs, LED matrix displays, and LCD panels.

Project 14
Multicolor Light Display

This project uses a high-brightness, three-color LED in combination with a rotary encoder. Turning the rotary encoder changes the color displayed by the LED.

The LED lamp is interesting because it has three LED lights in one four-pin package. The LED has a common-cathode arrangement, meaning that the negative connections of all three LEDs come out of one pin.

If you cannot find a four-pin RGB (red, green, blue) LED, you can use a six-pin device instead. Simply connect the separate anodes together, referring to the datasheet for the component.

Hardware

Figure 6-1 shows the schematic diagram for Project 14 and Figure 6-2 the breadboard layout.

Each LED has its own series resistor to limit the current to about 30 mA per LED.

The LED package has a slight flatness to one side. Pin 2 is the common cathode and is the longest pin.

The completed project is shown in Figure 6-3.

Each of the LEDs (red, green, and blue) is driven from a pulse-width modulation (PWM) output of the Arduino board so that by varying the output of each LED we can produce a full spectrum of visible-light colors.

The rotary encoder is connected in the same way as for Project 11. Rotating it changes the color, and pressing it will turn the LED on and off.

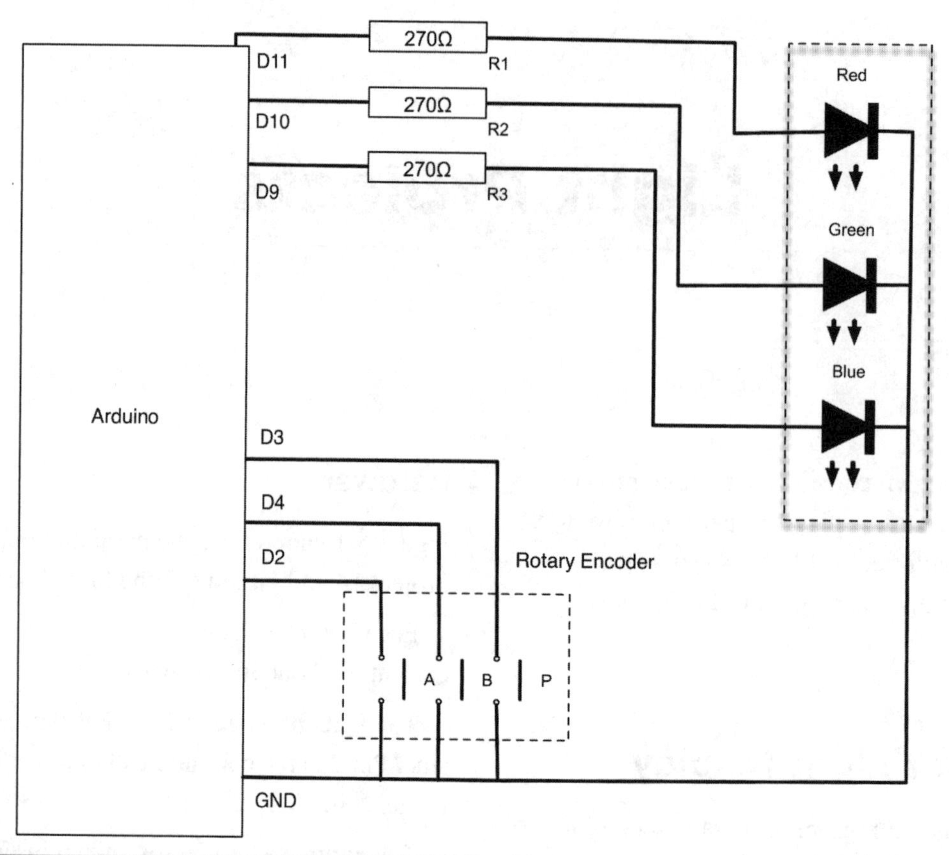

Figure 6-1 Schematic diagram for Project 14.

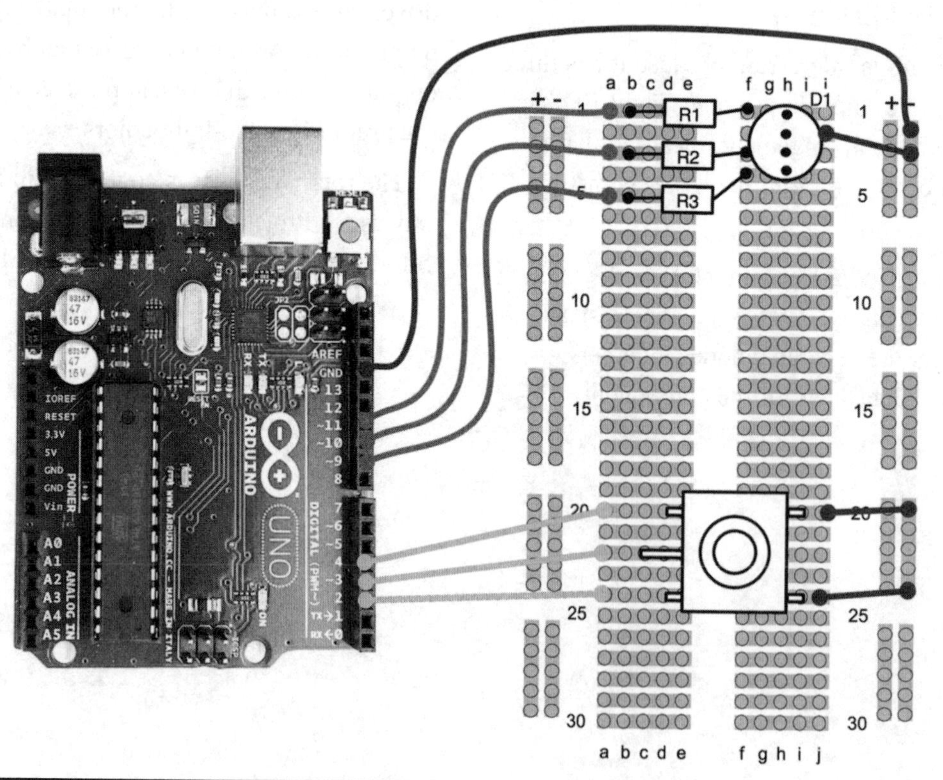

Figure 6-2 Breadboard layout for Project 14.

Figure 6-3 Project 14: multicolor light display.

Software

This sketch (Listing Project 14) uses an array to represent the different colors that will be displayed by the LED. Each of the elements of the array is a long 32-bit number. Three of the bytes of the long number are used to represent the red, green, and blue components of the color, which correspond to how brightly each of the red, green, or blue LED elements should be lit. The numbers in the array are shown in hexadecimal and correspond to the hex number format used to represent 24-bit colors on web pages. If there is a particular color that you want to try to create, find yourself a *web color chart* by typing "web color chart" into your favorite search engine. You can then look up the hex value for the color that you want.

LISTING PROJECT 14

```
int redPin = 9;
int greenPin = 10;
int bluePin = 11;
int aPin = 2;
int bPin = 4;
int buttonPin = 3;

boolean isOn = true;
int color = 0;
long colors[48]= {
```

(continued on next page)

LISTING PROJECT 14 (*continued*)

```
    0xFF2000, 0xFF4000, 0xFF6000, 0xFF8000, 0xFFA000, 0xFFC000, 0xFFE000, 0xFFFF00,
    0xE0FF00, 0xC0FF00, 0xA0FF00, 0x80FF00, 0x60FF00, 0x40FF00, 0x20FF00, 0x00FF00,
    0x00FF20, 0x00FF40, 0x00FF60, 0x00FF80, 0x00FFA0, 0x00FFC0, 0x00FFE0, 0x00FFFF,
    0x00E0FF, 0x00C0FF, 0x00A0FF, 0x0080FF, 0x0060FF, 0x0040FF, 0x0020FF, 0x0000FF,
    0x2000FF, 0x4000FF, 0x6000FF, 0x8000FF, 0xA000FF, 0xC000FF, 0xE000FF, 0xFF00FF,
    0xFF00E0, 0xFF00C0, 0xFF00A0, 0xFF0080, 0xFF0060, 0xFF0040, 0xFF0020, 0xFF0000
};

void setup()
{
  pinMode(aPin, INPUT);
  pinMode(bPin, INPUT_PULLUP);
  pinMode(buttonPin, INPUT_PULLUP);
  pinMode(redPin, OUTPUT);
  pinMode(greenPin, OUTPUT);
  pinMode(bluePin, OUTPUT);
}

void loop()
{
  if (digitalRead(buttonPin) == LOW)
  {
    isOn = ! isOn;
    delay(200); // de-bounce
  }
  if (isOn)
  {
    int change = getEncoderTurn();
    color = color + change;
    if (color < 0)
    {
      color = 47;
    }
    else if (color > 47)
    {
      color = 0;
    }
    setColor(colors[color]);
  }
  else
  {
  setColor(0);
  }
}

int getEncoderTurn()
{
```

LISTING PROJECT 14 (*continued*)

```
  // return -1, 0, or +1
  static int oldA = LOW;
  static int oldB = LOW;
  int result = 0;
  int newA = digitalRead(aPin);
  int newB = digitalRead(bPin);
  if (newA != oldA || newB != oldB)
  {
    // something has changed
    if (oldA == LOW && newA == HIGH)
    {
      result = -(oldB * 2 - 1);
    }
  }
  oldA = newA;
  oldB = newB;
  return result;
}
void setColor(long rgb)
{
  int red = rgb >> 16;
  int green = (rgb >> 8) & 0xFF;
  int blue = rgb & 0xFF;
  analogWrite(redPin, red);
  analogWrite(greenPin, green);
  analogWrite(bluePin, blue);
}
```

The 48 colors in the array are chosen from just such a table and are a range of colors more or less spanning the spectrum from red to violet.

Putting It All Together

Load the completed sketch for Project 14 from your Arduino Sketchbook and download it to the board (see Chapter 1).

Seven-Segment LEDs

There was a time when the height of fashion was an LED watch. This required the wearer to press a button on the watch for the time to magically appear as four bright-red digits. After a while, the inconvenience of having to use both limbs to tell the time overcame the novelty of a digital watch, and the Evil Genius went out and bought an LCD watch instead. This could only be read in bright sunlight.

Seven-segment LEDs (Figure 6-4) have largely been superseded by backlit LCD displays (see later in this chapter), but they do find uses from time to time. They also add that Evil Genius feel to a project.

Figure 6-5 shows the circuit for driving a single seven-segment display.

A single seven-segment LED usually does not have a great deal of use. Most projects will want two or four digits. When this is the case, we will not have enough digital output pins to drive each

Figure 6-4 Seven-segment LED display.

display separately, so the arrangement of Figure 6-6 is used.

Rather like our keyboard scanning, we are going to activate each display in turn and set the segments for that before moving on to the next digit. We do this so fast that the illusion of all displays being lit is created.

Each display potentially could draw the current for eight LEDs at once, which could amount to 160 mA (at 20 mA per LED)—far more than we can take from a digital output pin. For this reason, we use a transistor that is switched by a digital output to enable each display in turn.

The type of transistor we are using is called a *bipolar transistor*. It has three connections: emitter, base, and collector. When a current flows through the base of the transistor and out through the emitter, it allows a much greater current to flow through from the collector to the emitter. We have met this kind of transistor before in Project 4, where we used it to control the current to a high-power Luxeon LED.

We do not need to limit the current that flows through the collector to the emitter because this is already limited by the series resistors for the LEDs. However, we do need to limit the current flowing into the base. Most transistors will

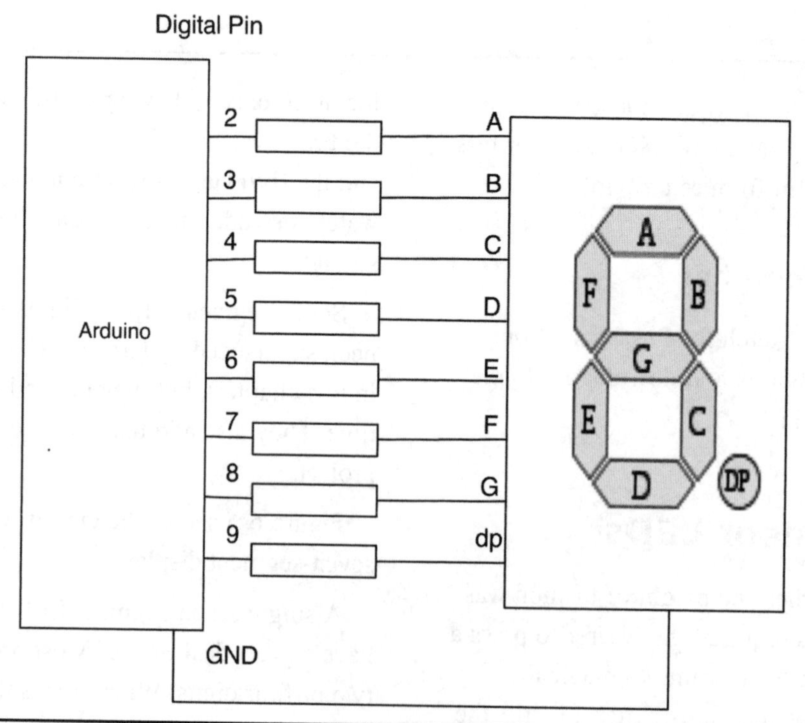

Figure 6-5 Arduino board driving a seven-segment LED.

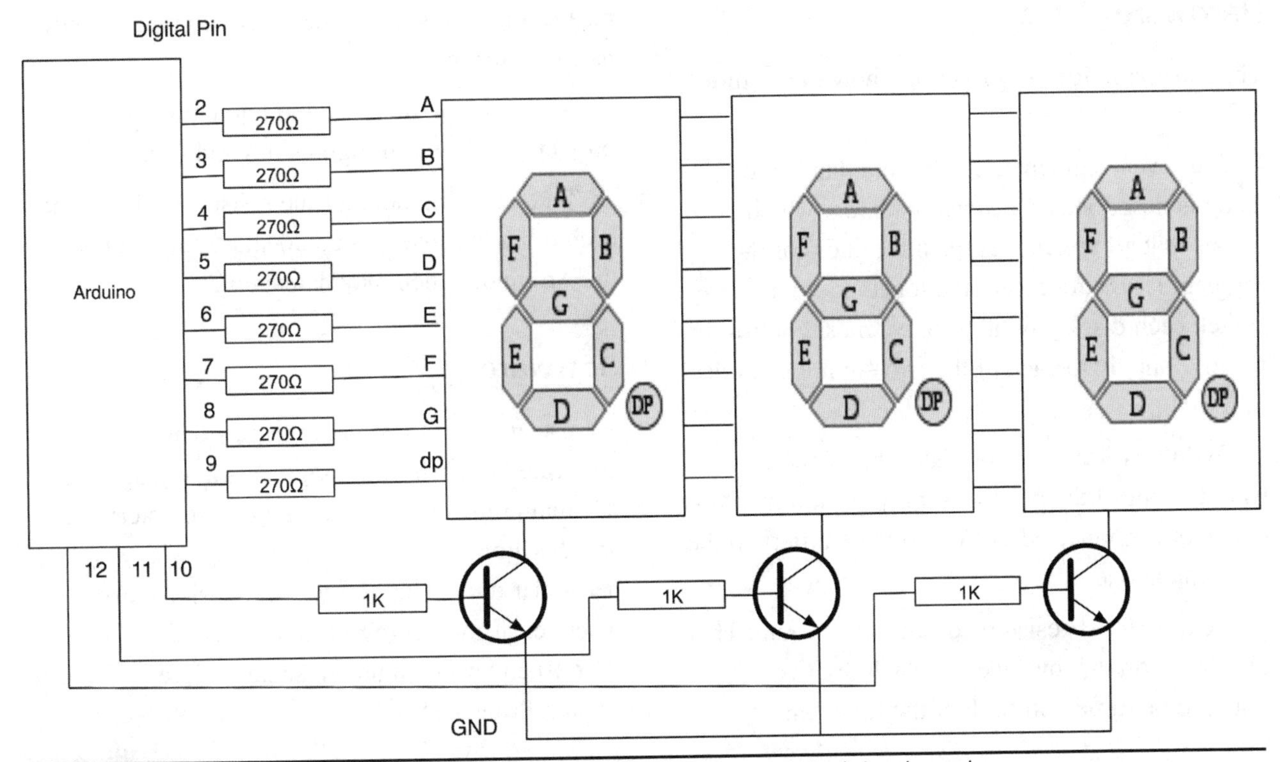

Figure 6-6 Driving more than one seven-segment LED from an Arduino board.

multiply the current by a factor of 100 or more, so we only need to allow about 2 mA to flow through the base to fully turn on the transistor.

Transistors have the interesting property that under normal use the voltage between base and emitter is a fairly constant 0.6V no matter how much current is flowing. So, if our Arduino pin supplies 5V, 0.6V of that will be across the base/emitter of the transistor, meaning that our resistor should have a value of about

$$R = V/I$$

$$R = 4.4/2\text{ mA} = 2.2\text{ k}\Omega$$

In actual fact it would be just fine if we let 4 mA flow because the digital output can cope with about 40 mA, so let's choose the nice standard resistor value of 1 kΩ, which will allow us to be sure that the transistor will act like a switch and always turn fully on or fully off.

Project 15
Seven-Segment LED Double Dice

In Project 9 we made a single dice using seven separate LEDs. In this project we will use two seven-segment LED displays to create a double dice.

COMPONENTS AND EQUIPMENT

	Description	Appendix
	Arduino Uno or Leonardo	m1/m2
D1	Two-digit, seven-segment LED display (common anode)	s8
R3-10	100 Ω, 0.25 W resistor	r2
R1, R2	1 kΩ, 0.25 W metal film resistor	r5
T1, T2	2N2222 transistor	s14
S1	Push switch	h3
	Solderless breadboard	h1
	Jumper wires	h2

Hardware

The schematic for this project is shown in Figure 6-7.

The seven-segment LED module that we are using is described as a *common anode*, which means that all the anodes (positive ends) of the segment LEDs are connected together. So, to switch each display on in turn, we must control the positive supply to each of the two common anodes in turn.

To do this, we use a transistor, but since we want to control the positive supply, each transitor's collector is connected to 5V and the emitters to the common anode.

We use 100 Ω resistors to limit the current. This may seem on the low side, but each digit is only going to be turned on for half the time, which means, on average, that the LED will receive only half the current.

The breadboard layout and photograph of the project are shown in Figures 6-8 and 6-9.

Take care that none of the resistor leads touches each other because this could short output pins on the Arduino, which may damage it.

Software

We use an array to contain the pins that are connected to each of the segments a to g and the decimal point. We also use an array to determine which segments should be lit to display any particular digit. This is a two-dimensional array, where each row represents a separate digit (0 to 9) and each column a segment (see Listing Project 15).

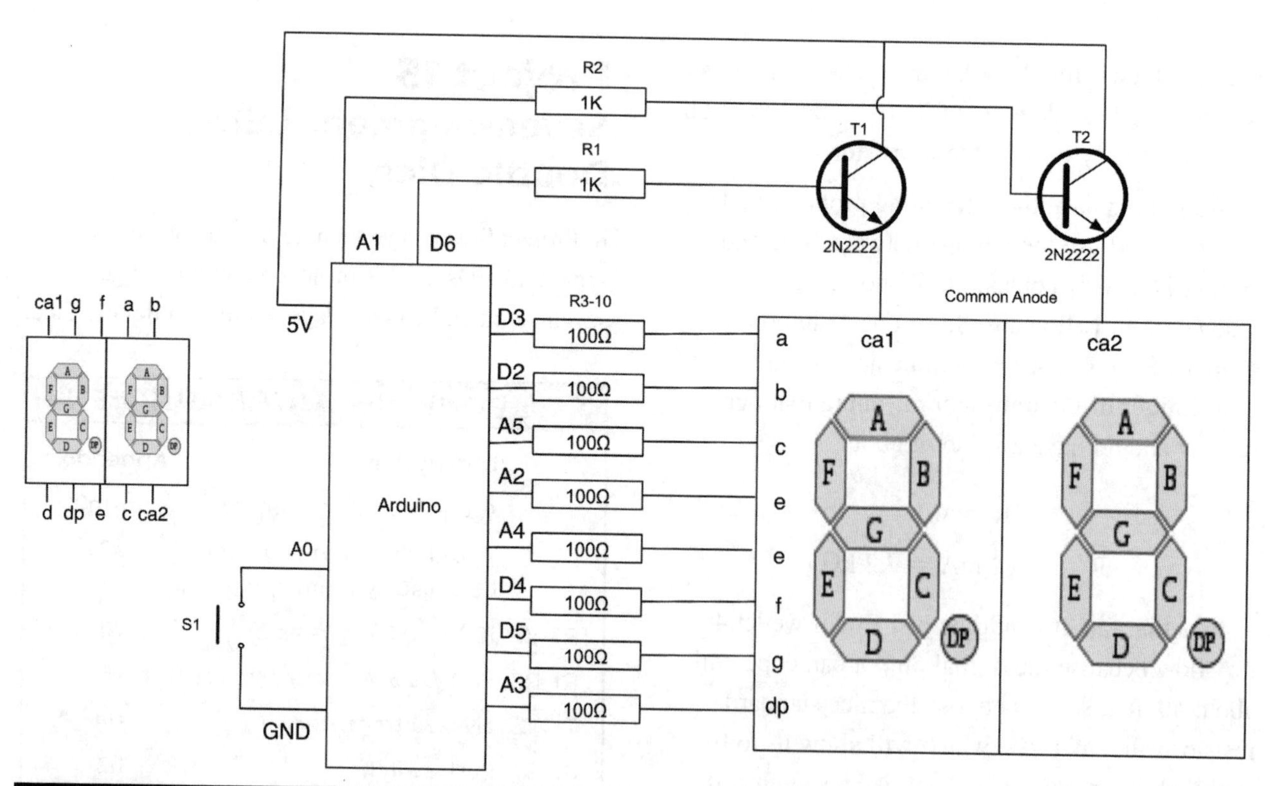

Figure 6-7 Schematic diagram for Project 15.

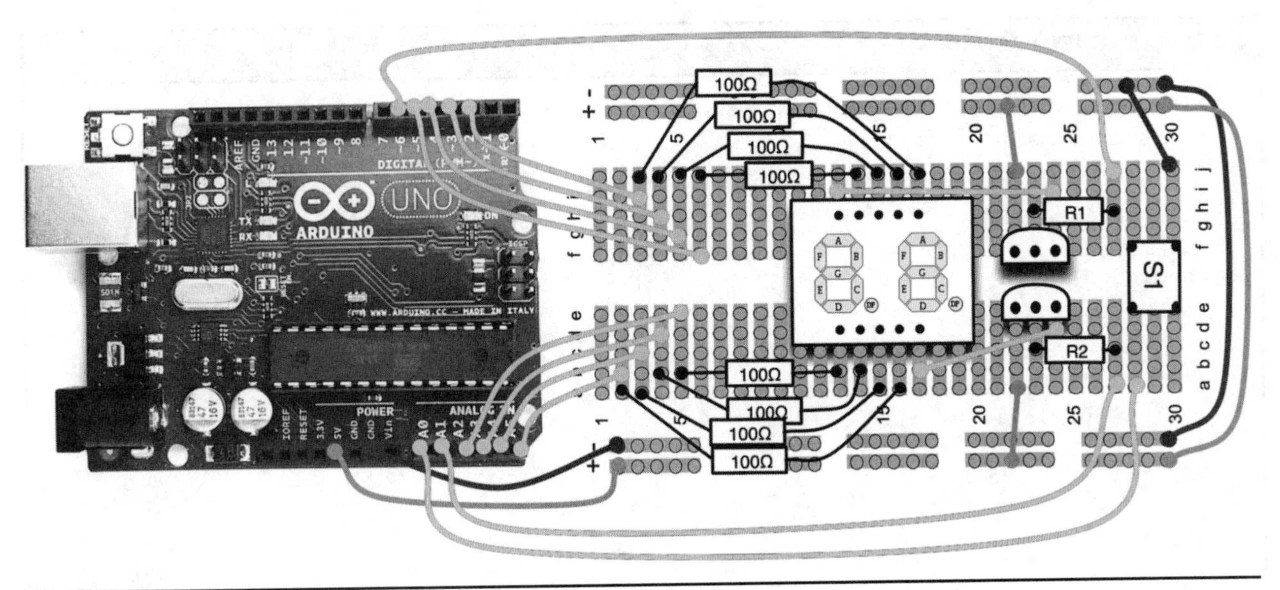

Figure 6-8 Breadboard layout for Project 15.

Figure 6-9 Double seven-segment LED dice.

LISTING PROJECT 15

```
int segmentPins[] = {3, 2, A5, A2, A4, 4, 5, A3};
int displayPins[] = {A1, 6};

int buttonPin = A0;

byte digits[10][8] = {
// a b c d e f g .
  { 1, 1, 1, 1, 1, 1, 0, 0}, // 0
  { 0, 1, 1, 0, 0, 0, 0, 0}, // 1
  { 1, 1, 0, 1, 1, 0, 1, 0}, // 2
  { 1, 1, 1, 1, 0, 0, 1, 0}, // 3
  { 0, 1, 1, 0, 0, 1, 1, 0}, // 4
  { 1, 0, 1, 1, 0, 1, 1, 0}, // 5
  { 1, 0, 1, 1, 1, 1, 1, 0}, // 6
  { 1, 1, 1, 0, 0, 0, 0, 0}, // 7
  { 1, 1, 1, 1, 1, 1, 1, 0}, // 8
  { 1, 1, 1, 1, 0, 1, 1, 0}  // 9
};

void setup()
{
  for (int i=0; i < 8; i++)
  {
    pinMode(segmentPins[i], OUTPUT);
  }
    pinMode(displayPins[0], OUTPUT);
    pinMode(displayPins[0], OUTPUT);
    pinMode(buttonPin, INPUT_PULLUP);
}

void loop()
{
  static int dice1;
  static int dice2;
  if (digitalRead(buttonPin) == LOW)
  {
    dice1 = random(1,7);
    dice2 = random(1,7);
  }
  updateDisplay(dice1, dice2);
}

void updateDisplay(int value1, int value2)
{
  digitalWrite(displayPins[0], HIGH);
  digitalWrite(displayPins[1], LOW);
  setSegments(value1);
```

LISTING PROJECT 15 *(continued)*

```
    delay(5);
    digitalWrite(displayPins[0], LOW);
    digitalWrite(displayPins[1], HIGH);
    setSegments(value2);
    delay(5);
}
void setSegments(int n)
{
    for (int i=0; i < 8; i++)
    {
        digitalWrite(segmentPins[i], ! digits[n][i]);
    }
}
```

To drive both displays, we have to turn each display on in turn, setting its segments appropriately. So our loop function must keep the values that are displayed in each display in separate variables: dice1 and dice2.

To throw the dice, we use the random function, and whenever the button is pressed, a new value will be set for dice1 and dice2. This means that the throw also will depend on how long the button is pressed, so we do not need to worry about seeding the random-number generator.

Putting It All Together

Load the completed sketch for Project 15 from your Arduino Sketchbook and download it to the board (see Chapter 1).

Project 16
LED Array

LED arrays are one of those components that just look like they would be useful to the Evil Genius. They consist of an array of LEDs (in this case 8 by 8). These devices can have just a single LED at each position; however, in the device that we are going to use, each of these LEDs is actually a pair of LEDs, one red and one green, positioned under a single lens so that they appear to be one dot. We can then light either one or both LEDs to make a red, green, or orange color.

The completed project is shown in Figure 6-10.

This project makes use of one of these devices and allows multicolor patterns to be displayed.

This project uses a very convenient module from Adafruit that includes a driver chip, which means that we need only two pins to control the LED matrix and two to provide power.

Figure 6-10 Project 16: LED array.

COMPONENTS AND EQUIPMENT	
Description	**Appendix**
Arduino Uno or Leonardo	m1/m2
8 by 8 LED bicolor I^2C module	m5
Solderless breadboard	h1
Jumper wires	h2

Hardware

The LED matrix module is supplied as a kit (Figure 6-11). This is very easy to assemble, and full instructions are provided on the Adafruit website. You will however need to do a little soldering.

The most important thing is to make sure that the LED matrix is soldered onto the board with the correct orientation. Once attached, it will be very difficult to change.

Figure 6-12 shows the schematic diagram for the project. The module uses a type of serial interface called I^2C (pronounced "I squared C"). This uses just two pins located after the GND and AREF pins. On a Leonardo, theses pins are labeled "SDA" and "SCL"; on a Uno, they are not labeled.

Another difference is that on the Leonardo these pins are dedicated for use as I^2C, whereas on the Uno they are also connected to A4 and A5. Thus, when using I^2C on an Arduino Uno, you cannot also use A4 and A5 as analog inputs.

If you have an old Arduino board without SDA and SCL sockets, you can use A4 and A5 instead.

With just four pins to connect, the breadboard layout is pretty trivial (Figure 6-13).

Software

The LED module needs two libraries to be installed. Both are available from the Adafruit website (http://learn.adafruit.com/adafruit-led-backpack/bi-color-8x8-matrix). The procedure for installing them is the same as installing the Keypad library back in Project 10.

When following the links to the libraries from the Adafruit website, look for the "Zip" option, which will download Zip files for the two libraries. The files are

- Adafruit-LED-Backpack-Library-master
- Adafruit-GFX-Library-master

Figure 6-11 The Adafruit bicolor LED matrix module kit.

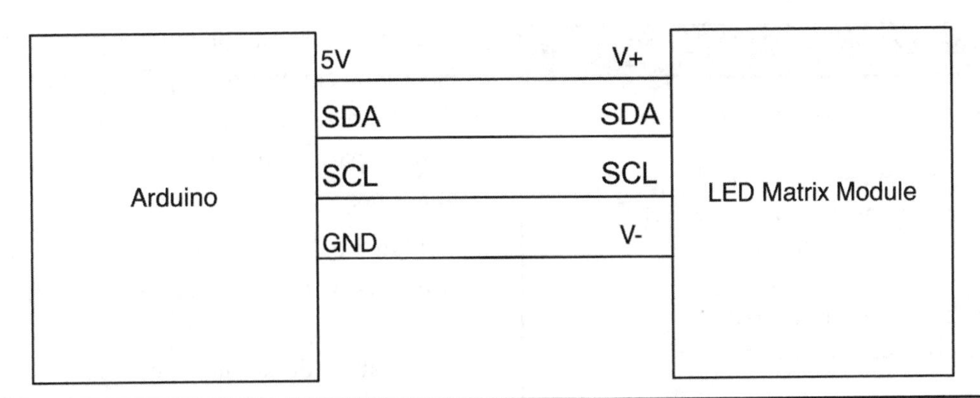

Figure 6-12 Schematic diagram for Project 16.

Extract these Zip archives into Documents/Arduino/libraries as you did with the Keypad library. You also will need to rename the folders to be "Adafruit_LEDBackpack" and "Adafruit_GFX."

Restart the Arduino IDE to pick up the new libraries and load the sketch Project 16_led_Matrix. You should see a nice colorful display.

The software for this project is quite short (Listing Project 16) and makes heavy use of the libraries.

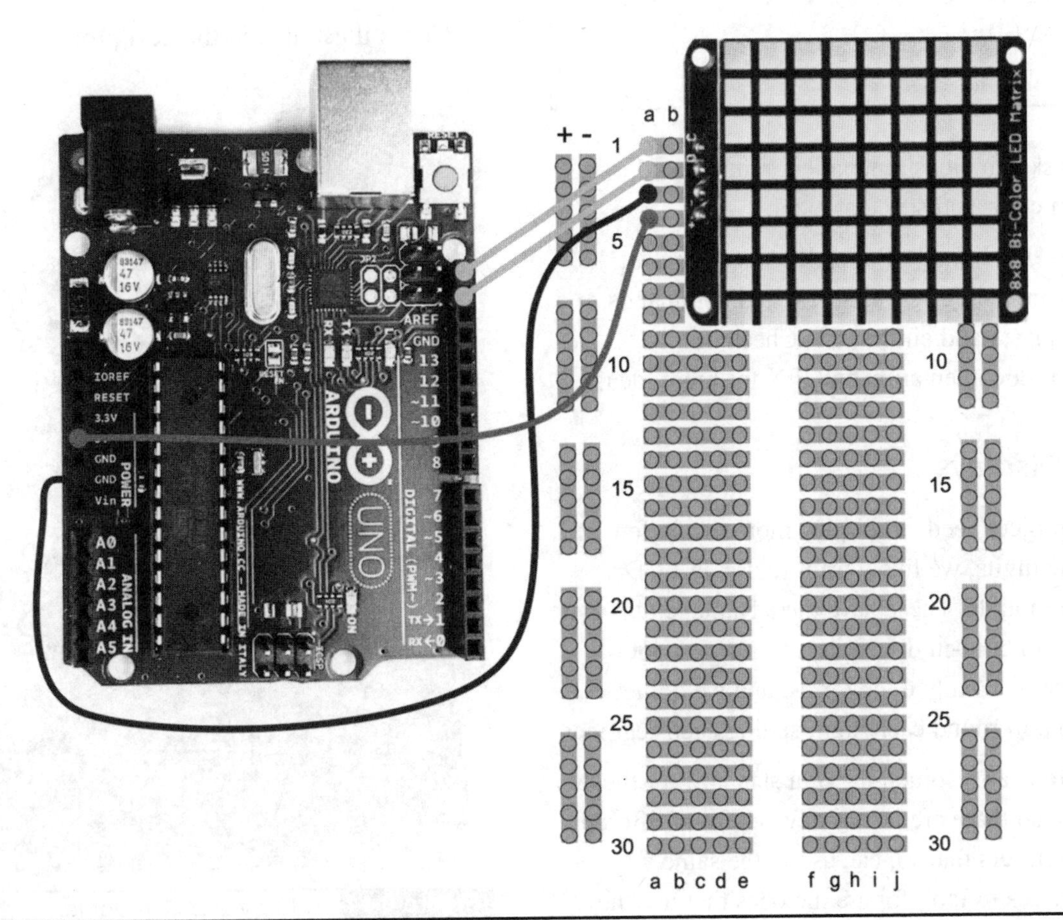

Figure 6-13 Breadboard layout for Project 16.

LISTING PROJECT 16

```
// Project 16 - LED MAtrix

#include <Wire.h>
#include "Adafruit_LEDBackpack.h"
#include "Adafruit_GFX.h"

Adafruit_BicolorMatrix matrix =
     Adafruit_BicolorMatrix();

void setup()
{
 matrix.begin(0x70);
}

void loop()
{
 uint16_t color = random(4);
 int x = random(8);
 int y = random(8);
 matrix.drawPixel(x, y, color);
 matrix.writeDisplay();
 delay(2);
}
```

The sketch picks random coordinates and a random color and sets that pixel.

The GFX library allows all sorts of special effects, including scrolling text and commands to draw squares and circles, etc. Check out the Adafruit documentation on GFX for more ideas.

LCD Displays

If our project needs to display more than a few numeric digits, we likely want to use an LCD display module. These have the advantage that they come with built-in driver electronics, so a lot of the work is already done for us, and we do not have to poll round each digit, setting each segment.

There is also something of a standard for these devices, so there are lots of devices from different manufacturers that we can use in the same way. The devices to look for are the ones that use the HD44780 driver chip.

LCD panels can be quite expensive from retail electronic component suppliers, but if you look on the Internet, they often can be bought for a few dollars, particularly if you are willing to buy a few at a time.

Figure 6-14 shows a module that can display two rows of 16 characters. Each character is made up of an array of 7 by 5 segments. So it is just as well that we do not have to drive each segment separately.

The display module includes a character set so that it knows which segments to turn on for any character. This means that we just have to tell it which character to display where on the display.

We need just seven digital outputs to drive the display. Four of these are data connections, and three control the flow of data. The actual details of what is sent to the LCD module can be ignored because there is a standard library that we can use.

This is illustrated in the next project.

Figure 6-14 A 16 by 2 LCD module.

Project 17
USB Message Board

This project will allow us to display a message on an LCD module from our computer. There is no reason why the LCD module needs to be right next to the computer, so you could use it on the end of a long USB lead to display messages remotely—next to an intercom at the door to the Evil Genius's lair, for example.

COMPONENTS AND EQUIPMENT

	Description	Appendix
	Arduino Uno or Leonardo	m1/m2
	LCD module (HD44780 controller)	m6
R1	10 kΩ trimpot	r11
	Strip of 0.1-inch header pins (at least 16)	h12
	Solderless breadboard	h1
	Jumper wires	h2

Hardware

The schematic diagram for the LCD display is shown in Figure 6-15 and the breadboard layout in Figure 6-16. As you can see, the only components required are the LCD module itself and the variable resistor to control the display's contrast.

The LCD module receives data 4 bits at a time through the connections D4–7. The LCD module also has connectors for D0–3, which are used only for transferring data 8 bits at a time. To reduce the number of pins required, we do not use these.

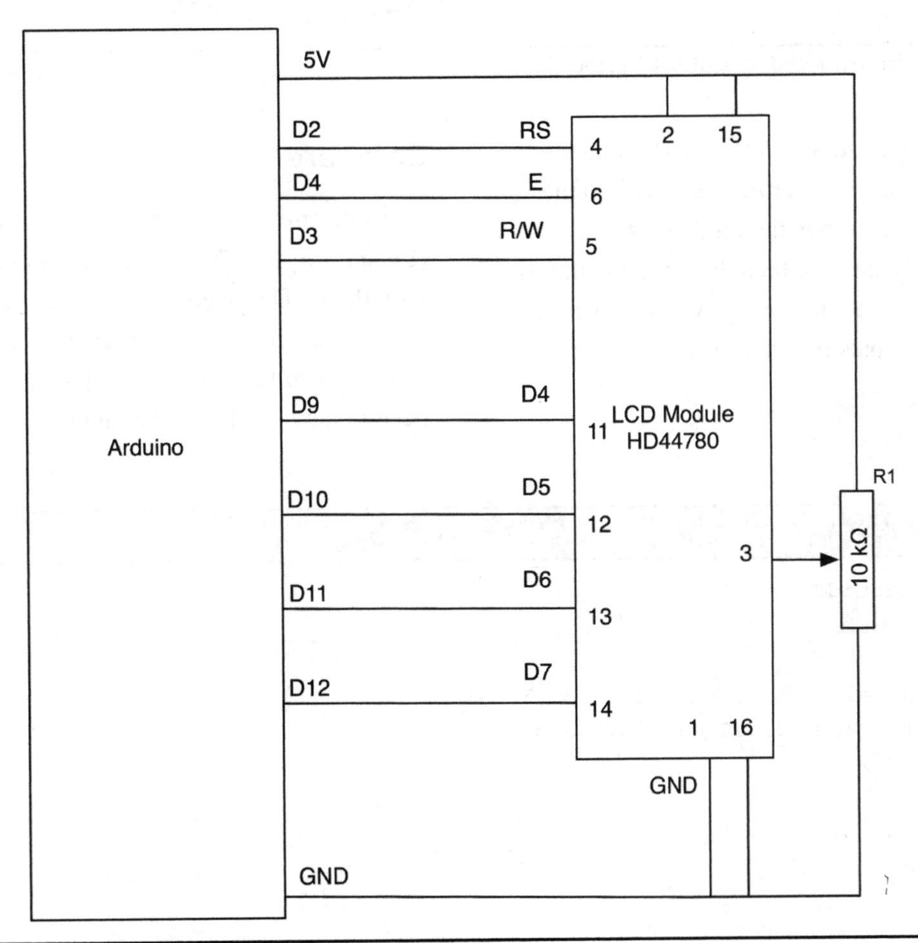

Figure 6-15 Schematic diagram for Project 17.

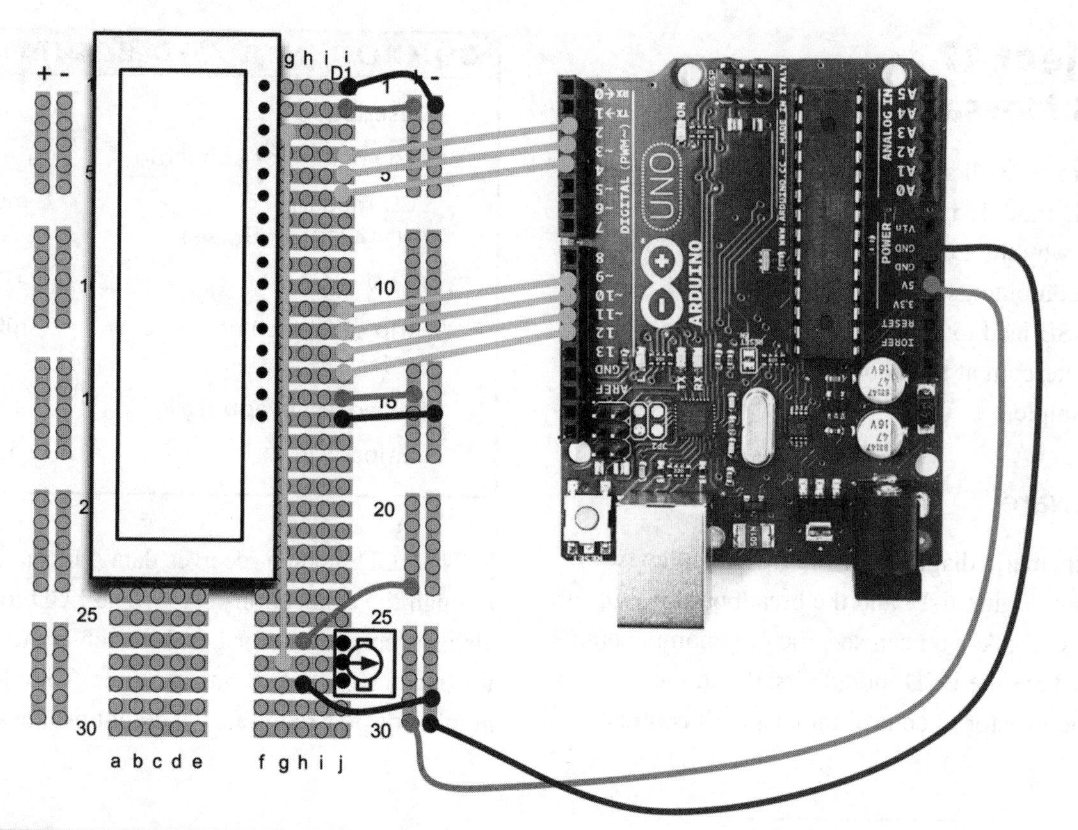

Figure 6-16 Breadboard layout for Project 17.

The easiest way to attach the LCD module to the breadboard is to solder header pins into the connector strip, and then the module can be plugged directly into the breadboard. Note that if you line pin 1 of the display up with row 1 of the breadboard, it makes it much easier to wire the project up.

Software

The software for this project is straightforward (Listing Project 17). All the work of communicating with the LCD module is taken care of by the LCD library. This library is included as part of the standard Arduino software installation, so we do not need to download or install anything special.

LISTING PROJECT 17

```
#include <LiquidCrystal.h>


//LiquidCrystal(rs, rw, enable, d4, d5, d6, d7)
LiquidCrystal lcd(12, 11, 10, 5, 4, 3, 2);

void setup()
{
  Serial.begin(9600);
  lcd.begin(2, 20);
  lcd.clear();
```

LISTING PROJECT 17 (*continued*)

```
  lcd.setCursor(0,0);
  lcd.print("Evil Genius");
  lcd.setCursor(0,1);
  lcd.print("Rules");
}

void loop()
{
  if (Serial.available())
  {
    char ch = Serial.read();
    if (ch == '#')
    {
      lcd.clear();
    }
    else if (ch == '/')
    {
      lcd.setCursor(0,1);
    }
    else
  {
      lcd.write(ch);
    }
  }
}
```

The loop function reads any input, and if it is a # character, it clears the display. If it is a / character, it moves to the second row; otherwise, it just displays the character that was sent.

Putting It All Together

Load the completed sketch for Project 17 from your Arduino Sketchbook and download it to the board (see Chapter 1).

You will probably need to turn the pot until the display contrast is just right.

We can now try out the project by opening the Serial Monitor and entering some text.

Later on in Project 22 we will be using the LCD panel again with a thermistor and rotary encoder to make a thermostat.

Summary

That's all for LED- and light-related projects. In Chapter 7 we will look at projects that use sound in one way or another.

CHAPTER 7

Sound Projects

AN ARDUINO BOARD CAN BE used to both generate sounds as an output and receive sounds as an input using a microphone. In this chapter we have various "musical instrument–type" projects as well as projects that process sound inputs.

Although not strictly a "sound" project, our first project is to create a simple oscilloscope so that we can view the waveform at an analog input.

Project 18
Oscilloscope

An *oscilloscope* is a device that allows you to see an electronic signal so that it appears as a waveform. A traditional oscilloscope works by amplifying a signal to control the position of a dot on the Y axis (vertical axis) of a cathode-ray tube while a time-base mechanism sweeps left to right on the X axis and then flips back when it reaches the end. The result will look something like Figure 7-1.

These days, cathode-ray tubes have largely been replaced by digital oscilloscopes that use LCD displays, but the principles remain the same.

This project reads values from the analog input and sends them over a USB cable to your computer. Rather than be received by the Serial Monitor, they are received by a little program that displays them in an oscilloscope-like manner. As the signal changes, so does the shape of the waveform.

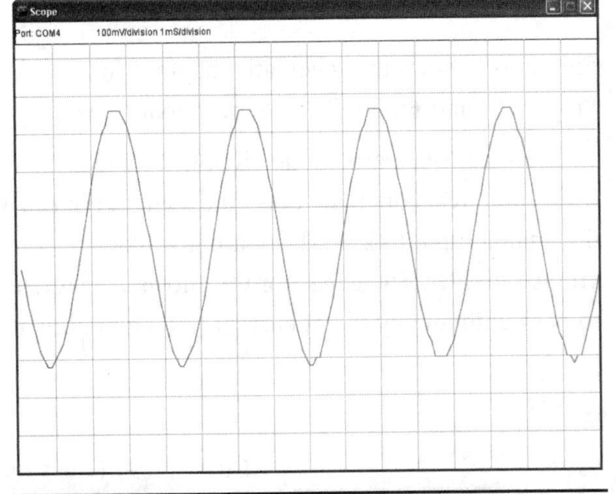

Figure 7-1 A 230 Hz sine wave on an oscilloscope.

Note that as oscilloscopes go, this one is not going to win any prizes for accuracy or speed, but it is kind of fun and will display waveforms up to about 1 kHz.

COMPONENTS AND EQUIPMENT

	Description	Appendix
	Arduino Uno or Leonardo	m1/m2
C1	220 nF capacitor	c2
C2, C3	100 μF electrolytic capacitor	c3
R1, R2	1 MΩ, 0.25 W resistor	r10
R3, R4	1 kΩ, 0.25 W resistor	r5
D1	5.1V zener diode	s13
	Solderless breadboard	h1
	Jumper wires	h2

This is the first time that we have used capacitors. C1 can be connected either way round; however, C2 and C3 are polarized and must be connected the correct way round or they are likely to be damaged. As with LEDs, on polarized capacitors, the positive lead (marked as the white rectangle on the schematic symbol) is longer than the negative lead. The negative lead also often has a minus sign (–) or a diamond shape next to the negative lead.

Hardware

Figure 7-2 shows the schematic diagram for Project 18 and Figure 7-3 the breadboard layout.

There are two parts to the circuit. R1 and R2 are high-value resistors that "bias" the signal going to the analog input to 2.5V. They are just like a voltage divider. The capacitor C1 allows the signal to pass without any direct current (DC) component to the signal (alternating current, or AC, mode in a traditional oscilloscope).

R3, R4, C2, and C3 just provide a stable reference voltage of 2.5V. The reason for this is so that our oscilloscope can display both positive and negative signals. So one terminal of our test lead is fixed at 2.5V; any signal on the other lead will be relative to that. A positive voltage will mean a value at the analog input of greater than 2.5V, and a negative value will mean a value at the analog input of less than 2.5V.

The diode D1 will protect the analog input from accidental overvoltage.

Figure 7-4 shows the completed oscilloscope.

Software

The sketch is short and simple (Listing Project 18). Its only purpose is to read the analog input and blast it out to the USB port as fast as possible.

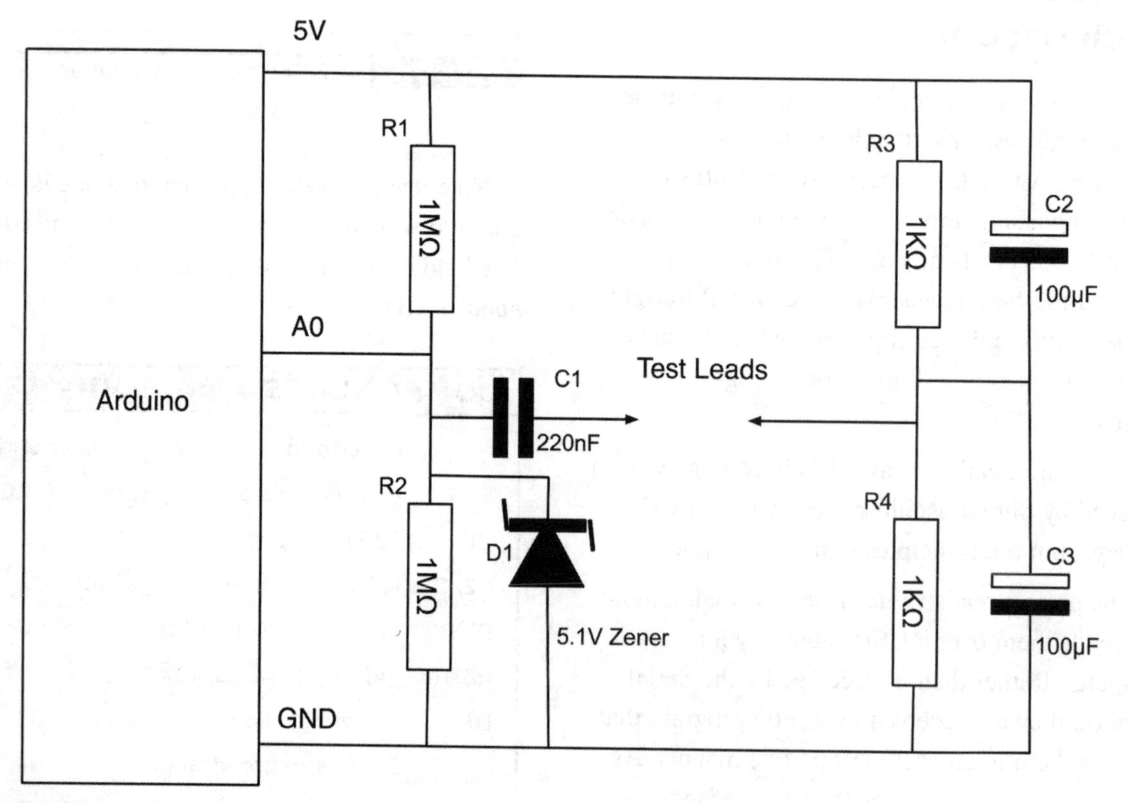

Figure 7-2 Schematic diagram for Project 18.

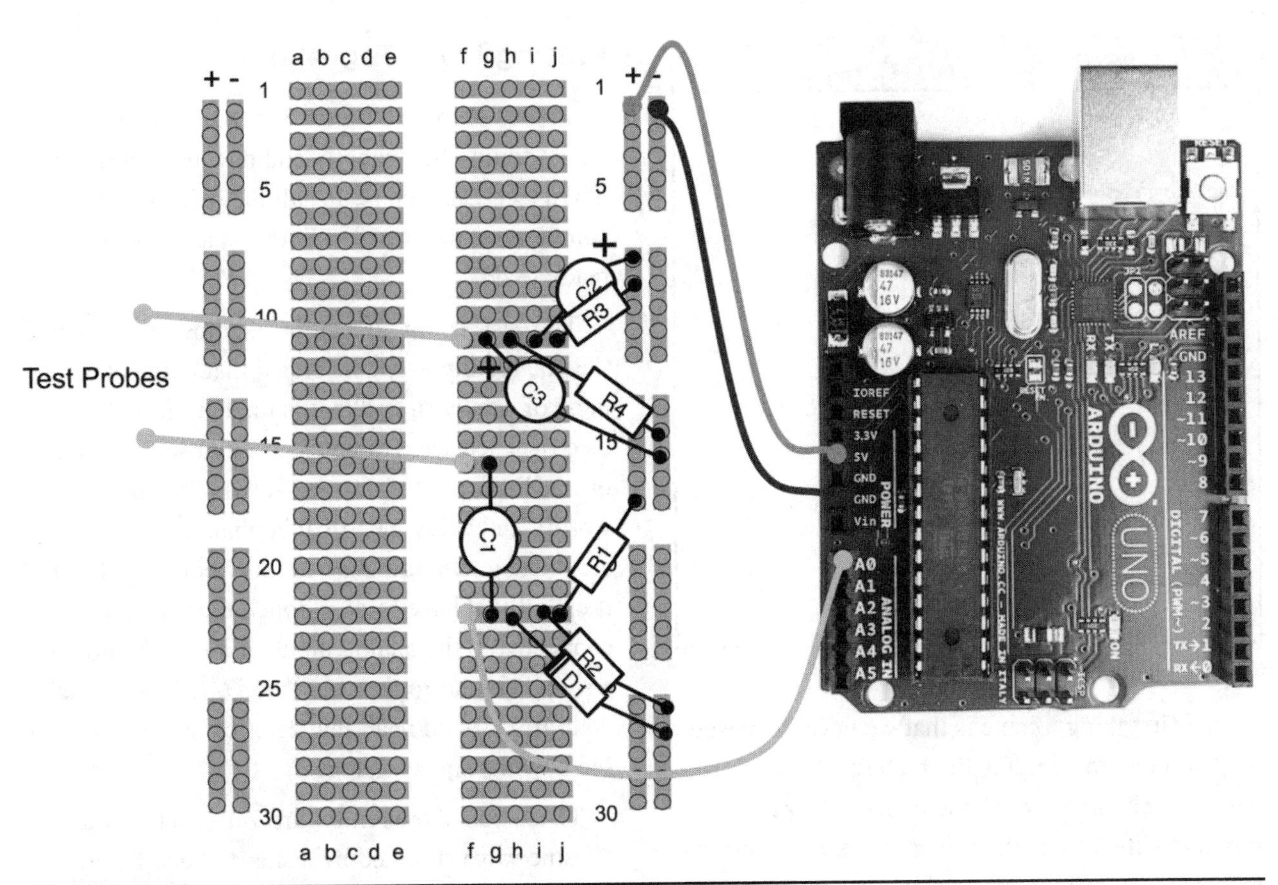

Figure 7-3 Breadboard layout for Project 18.

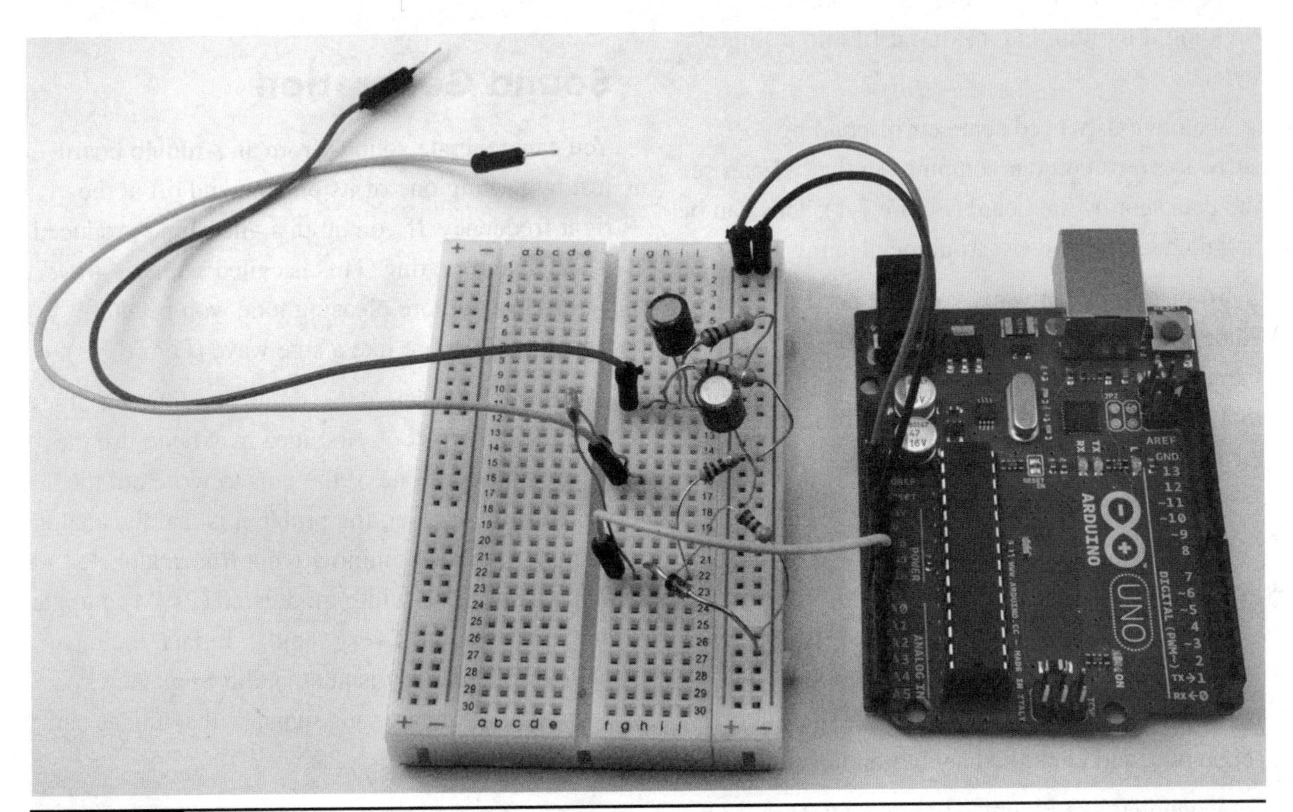

Figure 7-4 Project 18: oscilloscope.

LISTING PROJECT 18

```
// Project 18 - Oscilloscope

int analogPin = 0;

void setup()
{
  Serial.begin(115200);
}

void loop()
{
  int value = analogRead(analogPin);
  byte data = (value >> 2);
  Serial.write(data);
}
```

The first thing to note is that we have increased the baud rate to 115,200, the highest available. To get as much data through the connection as possible without resorting to complex compression techniques, we are going to shift our raw 10-bit value right 2 bits (>> 2); this has the effect of dividing it by four and making it fit into a single byte.

We obviously need some corresponding software to run on our computer so that we can see the data sent by the board (Figure 7-1). This can be downloaded from www.arduinoevilgenius.com.

To install the software, you first need to install some software called *Processing*. Processing is the natural partner for writing computer applications that communicate with an Arduino. In fact, the Arduino IDE is written in Processing.

Like the Arduino IDE, Processing is also available for Windows, Mac, and LINUX and can be downloaded from www.processing.org.

Once Processing is installed, run it. The similarities with the Arduino IDE will be immediately apparent. Now open the file scope.pde, and click the Play button to run it.

A window like Figure 7-1 should appear.

Putting It All Together

Load the completed sketch for Project 18 from your Arduino Sketchbook and download it to the board (see Chapter 1). Install the software for your computer as described previously, and you are ready to go.

The easiest way to test the oscilloscope is to use the one readily available signal that permeates most of our lives, and that is the hum from the electrical service. Home electricity oscillates at 50 or 60 Hz (depending on where you live in the world), and every electrical appliance emits electromagnetic radiation at this frequency. To pick it up, all you have to do is touch the test lead connected to the analog input, and you should see a signal similar to that of Figure 7-1. Try waving your arm around near any electrical equipment and see how the signal changes.

The signal shown in Figure 7-1 is actually a 215 Hz sine wave supplied by a smart phone function generator application.

Sound Generation

You can generate sounds from an Arduino board just by turning one of its pins on and off at the right frequency. If you do this, the sound produced is rough and grating. This is called a *square wave*. To produce a more pleasing tone, you need a signal that is more like a sine wave (Figure 7-5).

Generating a sine wave requires a little bit of thought and effort. A first idea may be to use the analog output of one of the pins to write out the waveform. However, the problem is that the analog outputs from an Arduino are not true analog outputs but pulse-width modulated (PWM) outputs that turn on and off very rapidly. In fact, their switching frequency is at an audio frequency, so without a lot of care, our signal will sound as bad as a square wave.

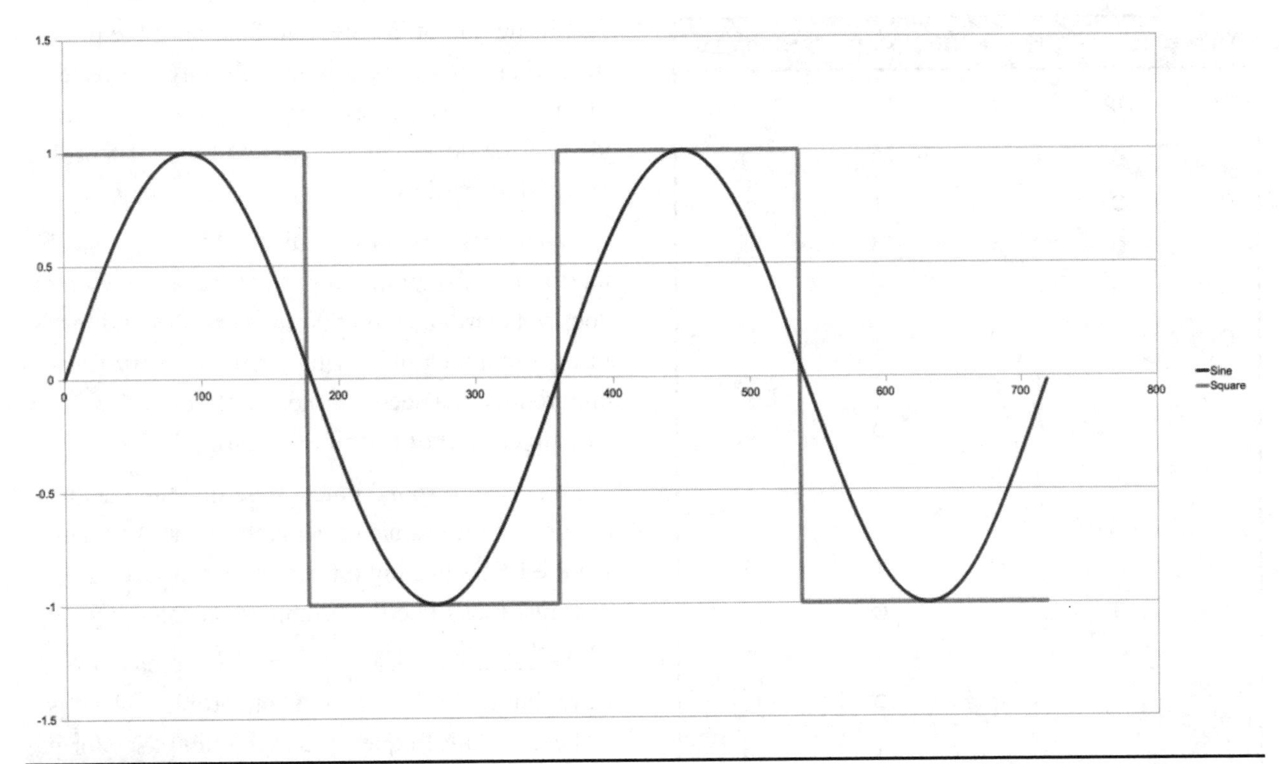

Figure 7-5 Square and sine waves.

A better way is to use a digital-to-analog converter (DAC). A DAC has a number of digital inputs and produces an output voltage proportional to the digital input value. Fortunately, it is easy to make a simple DAC—all you need are resistors.

Figure 7-6 shows a DAC made from what is called an R-2R resistor ladder.

It uses resistors of a value R and twice R, so R might be 5K and 2R 10K. Each of the digital inputs will be connected to an Arduino digital output. The four digits represent the four bits of the digital number. So this gives us 16 different analog outputs, as shown in Table 7-1.

Another way of generating a particular wave shape is to use the Arduino analogOutput command to generate the wave shape. This uses the technique of PWM that you first met back in Chapter 4 to control the brightness of LEDs.

Figure 7-7 shows the signal from a PWM pin on the Arduino.

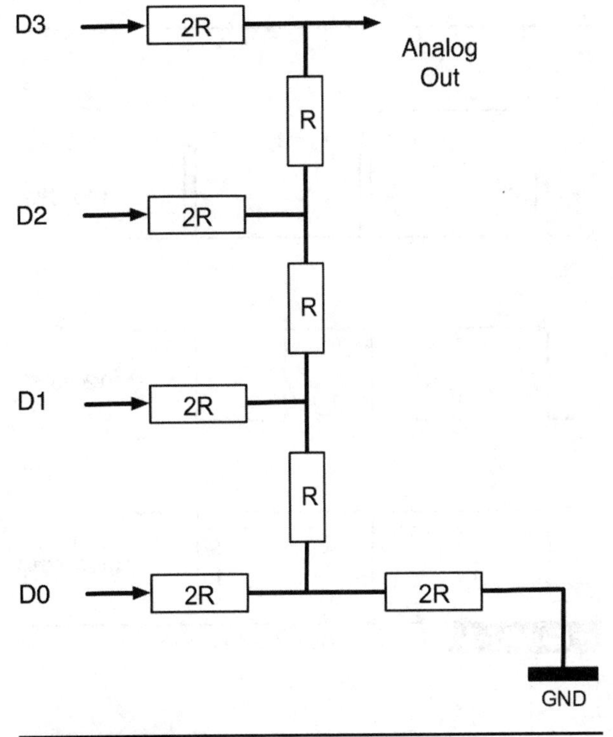

Figure 7-6 DAC using an R-2R ladder.

TABLE 7-1 Analog Output from Digital Inputs

D3	D2	D1	D0	Output
0	0	0	0	0
0	0	0	1	1
0	0	1	0	2
0	0	1	1	3
0	1	0	0	4
0	1	0	1	5
0	1	1	0	6
0	1	1	1	7
1	0	0	0	8
1	0	0	1	9
1	0	1	0	10
1	0	1	1	11
1	1	0	0	12
1	1	0	1	13
1	1	1	0	14
1	1	1	1	15

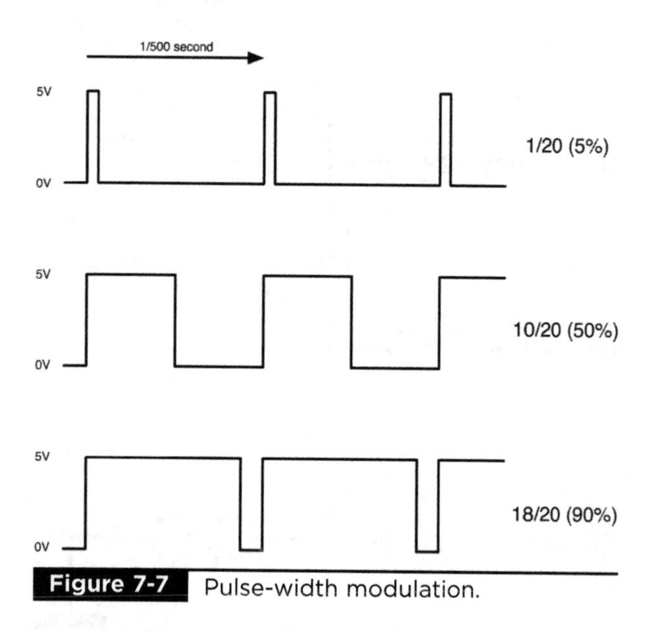

Figure 7-7 Pulse-width modulation.

The PWM pin is oscillating at about 500 times per second (hertz), with the relative amount of time that the pin is high varying with the value set in the analogWrite function. So, looking at Figure 7-7, if the output is only high for 5% of the time, then whatever we are driving will only receive 5% of full power. If, however, the output is at 5V for 90% of the time, then the load will get 90% of the power delivered to it.

When driving motors with PWM, the physical inertia of the spinning motor means that the motor does not start and stop 500 times per second but is just given a kick of varying strengths every five-hundredths of a second. The net effect of this is smooth control of the motor speed.

LEDs can respond much more quickly than a motor, but the visible effect is the same. We cannot see the LEDs turning on and off at that speed, so to us it just looks like the brightness is changing.

We can use this same technique to create a sine wave, but to do this, there is one problem. That is that the default frequency that Arduino uses for its PWM pulses is around 500 Hz, which is well within the range of audible frequencies. Fortunately, we can change this frequency in our sketch, making it much higher and outside the range of our hearing.

Figure 7-8 shows two traces from an oscilloscope of a 254 Hz sine wave being generated by writing out successive values from an array.

The array contains a series of values that, when used to set the value of analogWrite, one after the other produce the effect of a sine wave.

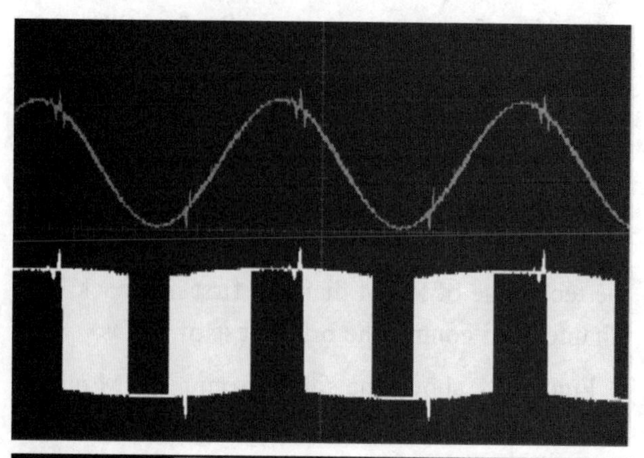

Figure 7-8 Oscilloscope trace for sine wave generation.

The bottom trace shows the raw PWM signal, with the pulses bunched up for the peaks and troughs of the sine wave and more spread out for the parts of the sine wave in the middle. The top trace shows that same signal after it has been passed through a low-pass filter that chops off the high PWM frequency (63 kHz), leaving us with quite a well-shaped sine wave.

Project 19
Tune Player

This project will play a series of musical notes through a miniature loudspeaker using PWM to approximate a sine wave.

If you can get a miniature loudspeaker with leads for soldering to a printed circuit board (PCB), then this can be plugged directly into the breadboard. If not, you will either have to solder short lengths of solid-core wire to the terminals or, if you do not have access to a soldering iron, carefully twist some wires round the terminals.

COMPONENTS AND EQUIPMENT

	Description	Appendix
	Arduino Uno or Leonardo	m1/m2
C1	100 nF nonpolarized capacitor	c1
C2	100 μF, 16V electrolytic capacitor	c3
R1	470 Ω, 0.25 W resistor	r4
R2	10 kΩ trimpot	r11
IC1	TDA7052 1 W audio amplifier	s23
	Miniature 8 Ω loudspeaker	h14
	Solderless breadboard	h1
	Jumper wires	h2

Hardware

To try to keep the number of components to a minimum, we have used an integrated circuit (IC) to amplify the signal and drive the loudspeaker. The TDA7052 IC provides 1 W of power output in an easy-to-use little 8-pin chip.

Figure 7-9 shows the schematic diagram for Project 19, and the breadboard layout is shown in Figure 7-10.

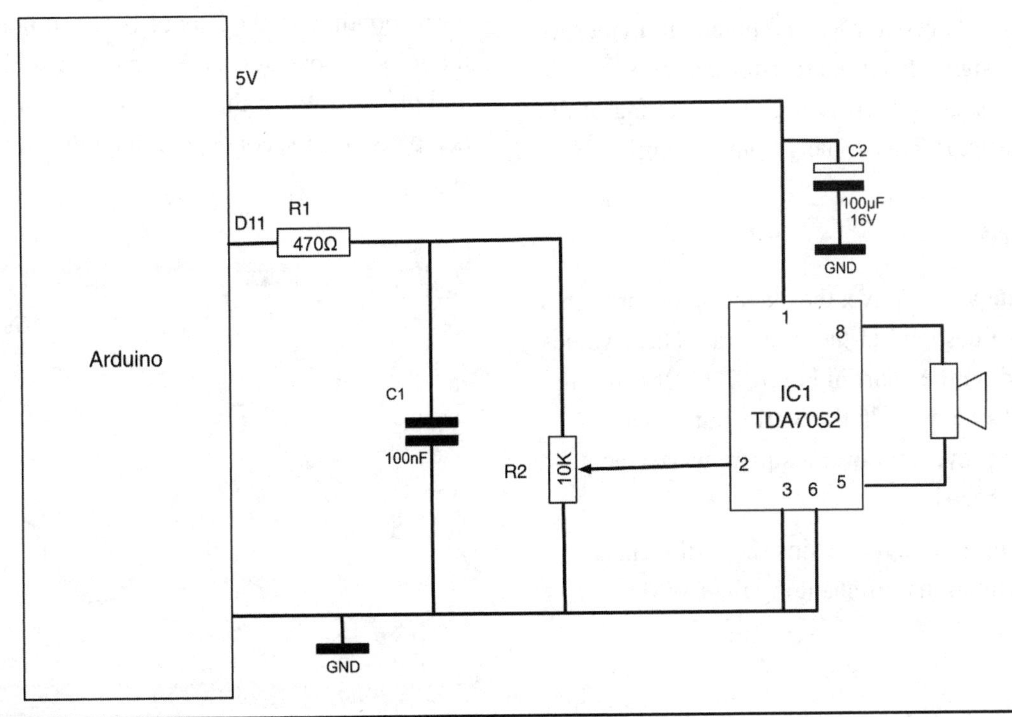

Figure 7-9 Schematic diagram for Project 19.

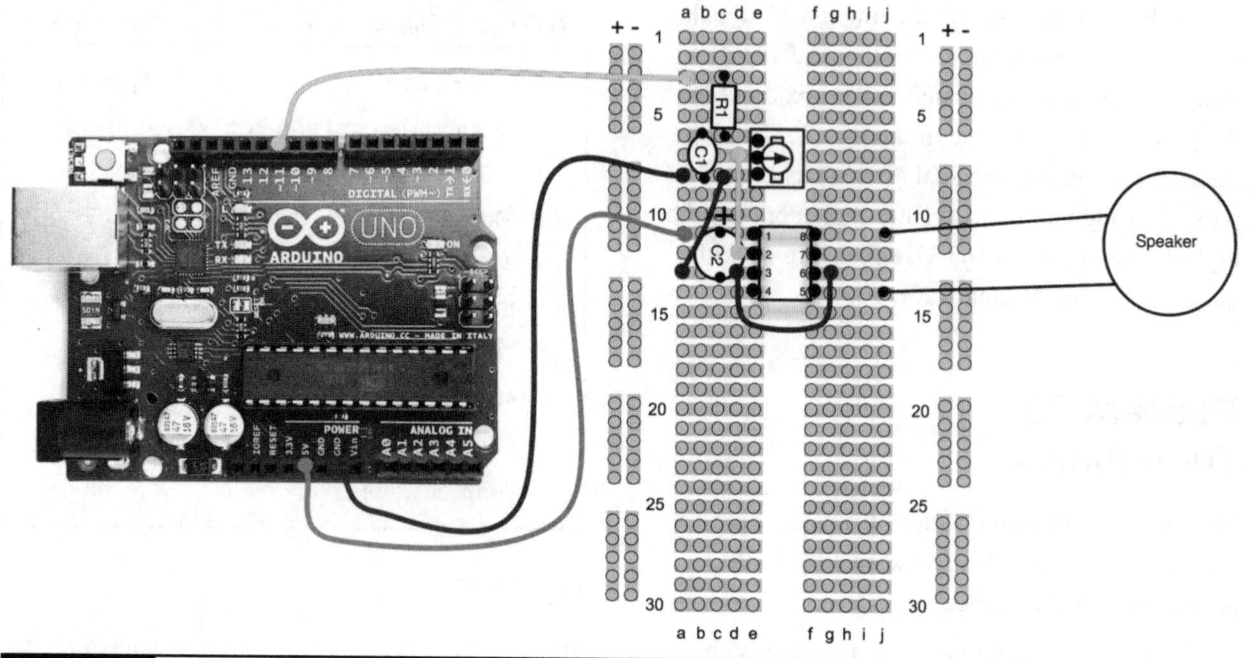

Figure 7-10 Breadboard layout for Project 19.

R1 and C1 together make a low-pass filter that will filter out the high-frequency PWM noise before it is passed on to the amplifier chip.

C2 is used as a decoupling capacitor that shunts any noise on the power lines to ground. This should be positioned as close as possible to IC1.

The variable resistor R2 is a potential divider to reduce the signal from the resistor ladder by at least a factor of 10, depending on the setting of the variable resistor. This is the volume control.

Software

To generate a sine wave, the sketch steps through a series of values held in the sine array. These values are plotted on the chart in Figure 7-11. It is not the smoothest sine wave in the world, but it is a definite improvement over a square wave (see Listing Project 19).

The setup function contains the Evil Genius's magic incantations for changing the PWM frequency.

The playNote function is the key to generating the note. The pitch of the note generated is controlled by the delay after each step of the signal within the playSine function that playNote calls.

Tunes are played from an array of characters, each character corresponding to a note and a space corresponding to the silence between notes. The main loop looks at each letter in the song variable and plays it. When the whole song is played, there is a pause of 5 seconds, and then the song begins again.

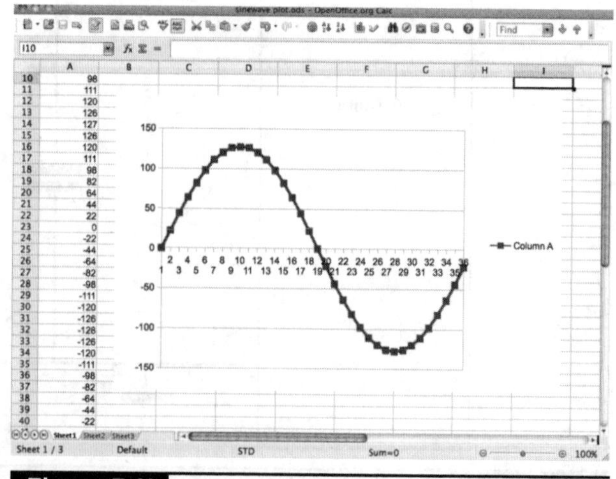

Figure 7-11 A plot of the sine array.

LISTING PROJECT 19

```
// Project 19 Tune Player

int soundPin = 11;

byte sine[] = {0, 22, 44, 64, 82, 98, 111, 120, 126, 127,
126, 120, 111, 98, 82, 64, 44, 22, 0, -22, -44, -64, -82,
-98, -111, -120, -126, -128, -126, -120, -111, -98, -82,
-64, -44, -22};

int toneDurations[] = {120, 105, 98, 89, 78, 74, 62};

char* song = "e e ee e e ee e g c d eeee f f f f f e e e e d d e dd gg e e ee e e
ee e g c d eeee f f f f f e e e g g f d cccc";

void setup()
{
  // change PWM frequency to 63kHz
  cli(); //disable interrupts while registers are configured
  bitSet(TCCR2A, WGM20);
  bitSet(TCCR2A, WGM21);      //set Timer2 to fast PWM mode (doubles PWM frequency)
  bitSet(TCCR2B, CS20);
  bitClear(TCCR2B, CS21);
  bitClear(TCCR2B, CS22);
  sei();                      //enable interrupts now that registers have been set
  pinMode(soundPin, OUTPUT);
}

void loop()
{
  int i = 0;
  char ch = song[0];
  while (ch != 0)
  {
    if (ch == ' ')
    {
      delay(75);
    }
    else if (ch >= 'a' and ch <= 'g')
    {
      playNote(toneDurations[ch - 'a']);
    }
    i++;
    ch = song[i];
  }

  delay(5000);
```

(continued on next page)

LISTING PROJECT 19 (*continued*)

```
}

void playNote(int pitchDelay)
{
  long numCycles = 5000 / pitchDelay;
  for (int c = 0; c < numCycles; c++)
  {
    playSine(pitchDelay);
  }
}

void playSine(int period)
{
  for( int i = 0; i < 36; i++)
  {
    analogWrite(soundPin, sine[i] + 128);
    delayMicroseconds(period);
  }
}
```

The Evil Genius will find this project useful for inflicting discomfort on his or her enemies.

Putting It All Together

Load the completed sketch for Project 19 from your Arduino Sketchbook and download it to the board (see Chapter 1).

You might like to change the tune played from "Jingle Bells." To do this, just comment out the line starting with char* song = by putting // in front of it, and then define your own array.

For a longer-duration note, just repeat the note letter without putting a space in between.

You will have noticed that the quality is not great. It is still a lot less nasty than using a square wave but is a long way from the tunefulness of a real musical instrument, where each note has an "envelope" in which the amplitude (volume) of the note varies with the note as it is played.

Project 20
Light Harp

This project is really an adaptation of Project 19 that uses two light sensors (LDRs): one that controls the pitch of the sound the other that

COMPONENTS AND EQUIPMENT

	Description	Appendix
	Arduino Uno or Leonardo	m1/m2
C1	100 nF unpolarized capacitor	c1
C2	100 µF, 16V electrolytic capacitor	c3
R1	470 Ω, 0.25 W resistor	r4
R2, R3	1 kΩ, 0.25 W resistor	7r5
R6	10 kΩ trimpot	r11
R4, R5	LDR	r13
IC1	TDA7052 1W audio amplifier	s23
	Miniature 8 Ω loudspeaker	h14
	Solderless breadboard	h1
	Jumper wires	h2

controls the volume. This is inspired by the Theremin musical instrument that is played by mysteriously waving your hands about between two antennas. In actual fact, this project produces a sound more like a bagpipe than a harp, but it is quite fun.

Hardware

Figures 7-12 and 7-13 show the schematic diagram and breadboard layout for the project, and you can see the final project in Figure 7-14.

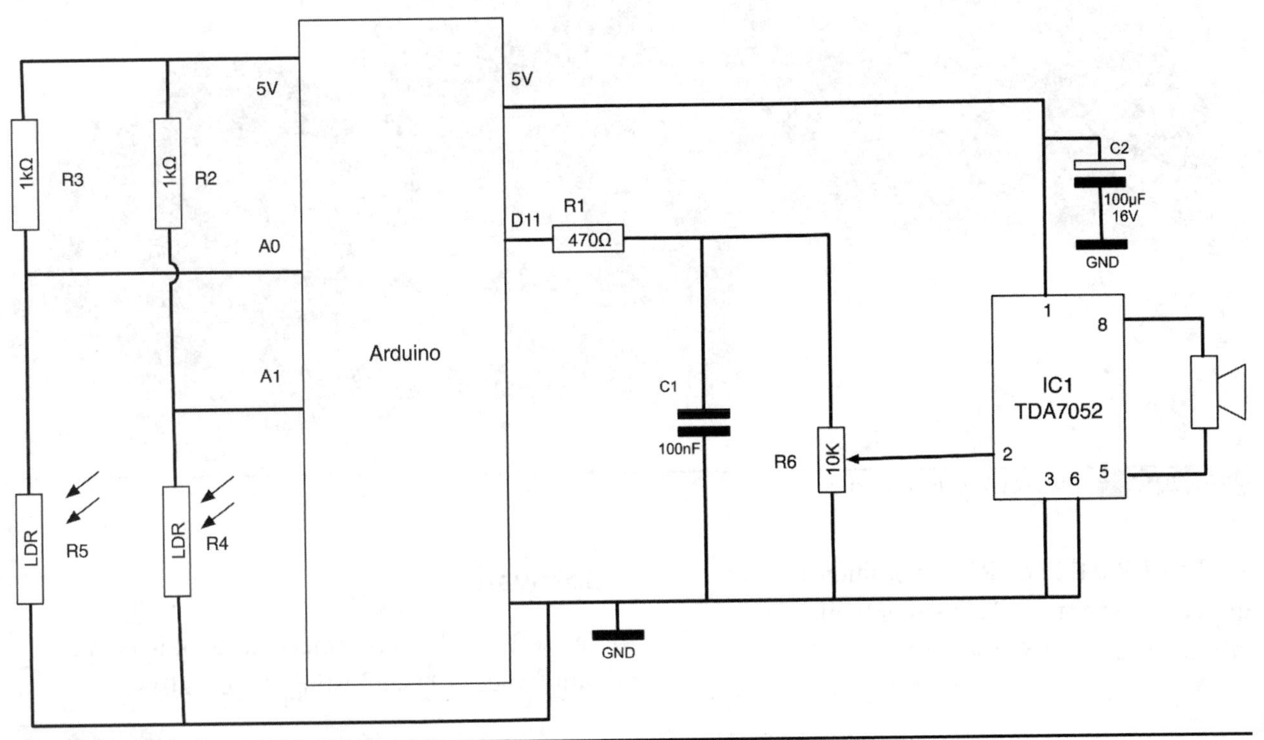

Figure 7-12 Schematic diagram for Project 20.

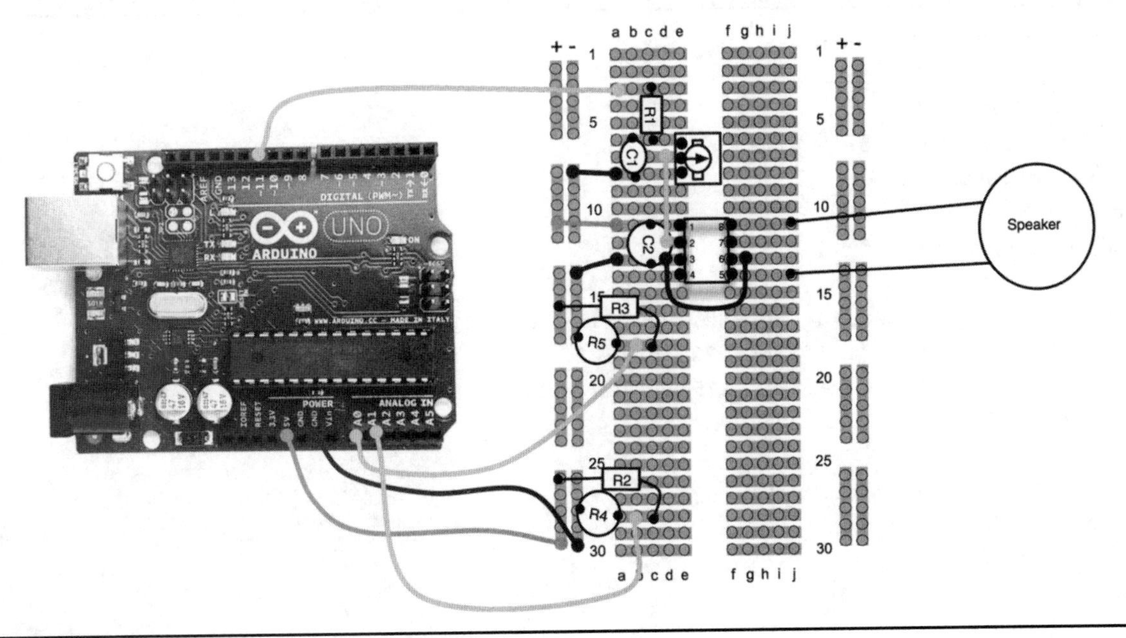

Figure 7-13 Breadboard layout for Project 20.

Figure 7-14 Project 20: light harp.

The LDRs, R4 and R5, are positioned away from each other to make it easier to play the instrument with two hands.

Software

The software for this project has a lot in common with Project 19 (see Listing Project 20).

LISTING PROJECT 20

```
// Project 20 Light Harp

int soundPin = 11;
int pitchInputPin = 0;
int volumeInputPin = 1;
int ldrDim = 400;
int ldrBright = 800;

byte sine[] = {0, 22, 44, 64, 82, 98, 111, 120, 126, 127,
126, 120, 111, 98, 82, 64, 44, 22, 0, -22, -44, -64, -82,
-98, -111, -120, -126, -128, -126, -120, -111, -98, -82,
-64, -44, -22};

long lastCheckTime = millis();
int pitchDelay;
int volume;

void setup()
```

LISTING PROJECT 20 (*continued*)

```
{
  // change PWM frequency to 63kHz
  cli();                        //disable interrupts while registers are configured
  bitSet(TCCR2A, WGM20);
  bitSet(TCCR2A, WGM21);        //set Timer2 to fast PWM mode (doubles PWM frequency)
  bitSet(TCCR2B, CS20);
  bitClear(TCCR2B, CS21);
  bitClear(TCCR2B, CS22);
  sei();                        //enable interrupts now that registers have been set
  pinMode(soundPin, OUTPUT);
}

void loop()
{
  long now = millis();
  if (now > lastCheckTime + 20L)
  {
    pitchDelay = map(analogRead(pitchInputPin), ldrDim, ldrBright, 10, 30);
    volume = map(analogRead(volumeInputPin), ldrDim, ldrBright, 1, 4);
    lastCheckTime = now;
  }

  playSine(pitchDelay, volume);
}

void playSine(int period, int volume)
{
  for( int i = 0; i < 36; i++)
  {
    analogWrite(soundPin, (sine[i] / volume) + 128);
    delayMicroseconds(period);
  }
}
```

The main differences are that the period passed to playSine is set by the value of the analog input 0. This is then scaled to the right range using the map function. Similarly, the volume voltage is set by reading the value of analog input 1, scaling it using map, and then using it to scale the values from the sine array before outputting them.

LDRs have different ranges of resistance. So you may find that you need to tweek the values of the variables ldrDim and ldrBright to get better ranges of pitch and volume.

Putting It All Together

Load the completed sketch for Project 20 from your Arduino Sketchbook and download it to the board (see Chapter 1).

To play the "instrument," use your right hand over one LDR to control the volume of the sound and your left hand over the other LDR to control the pitch. Interesting effects can be achieved by waving your hands over the LDRs.

Project 21
VU Meter

This project (shown in Figure 7-15) uses LEDs to display the volume of noise picked up by a microphone. It uses an array of LEDs built into a dual-in-line (DIL) package.

The push button toggles the mode of the VU meter. In normal mode, the bar graph just flickers up and down with the volume of sound. In maximum mode, the bar graph registers the maximum value and lights that LED, so the sound level gradually pushes it up.

Hardware

The schematic diagram for this project is shown in Figure 7-16. The bar-graph LED package has separate connections for each LED. These are each driven through a current-limiting resistor.

Figure 7-15 Project 21: VU meter.

COMPONENTS AND EQUIPMENT

	Description	Appendix
	Arduino Uno or Leonardo	m1/m2
R1, R3	10 kΩ, 0.25 W resistor	r6
R2	100 kΩ, 0.25 W resistor	r8
R4-13	270 Ω, 0.25 W resistor	r3
C1	100 nF capacitor	c1
T1	2N2222 transistor	s14
	10-segment bar-graph display	s9
S1	Push button to make switch	h3
	Electret microphone	h15
	Solderless breadboard	h1
	Jumper wires	h2

The microphone will not produce a strong enough signal on its own to drive the analog input. So, to boost the signal, we use a simple single-transistor amplifier. We use a standard arrangement called *collector-feedback bias*, where a proportion of the voltage at the collector is used to bias the transistor on so that it amplifies in a loosely linear way rather than just harshly switching on and off.

The breadboard layout is shown in Figure 7-17. With so many LEDs, a lot of wires are required. Make sure that the bar-graph LED module has the negative LED connections to the left of the breadboard as it appears in Figure 7-17. If it is not labeled, then test it out using one of the 270 Ω resistors and the 5V supply from the Arduino.

Software

The sketch for this project (Listing Project 21) uses an array of LED pins to shorten the setup function. This is also used in the loop function, where we iterate over each LED, deciding whether to turn it on or off.

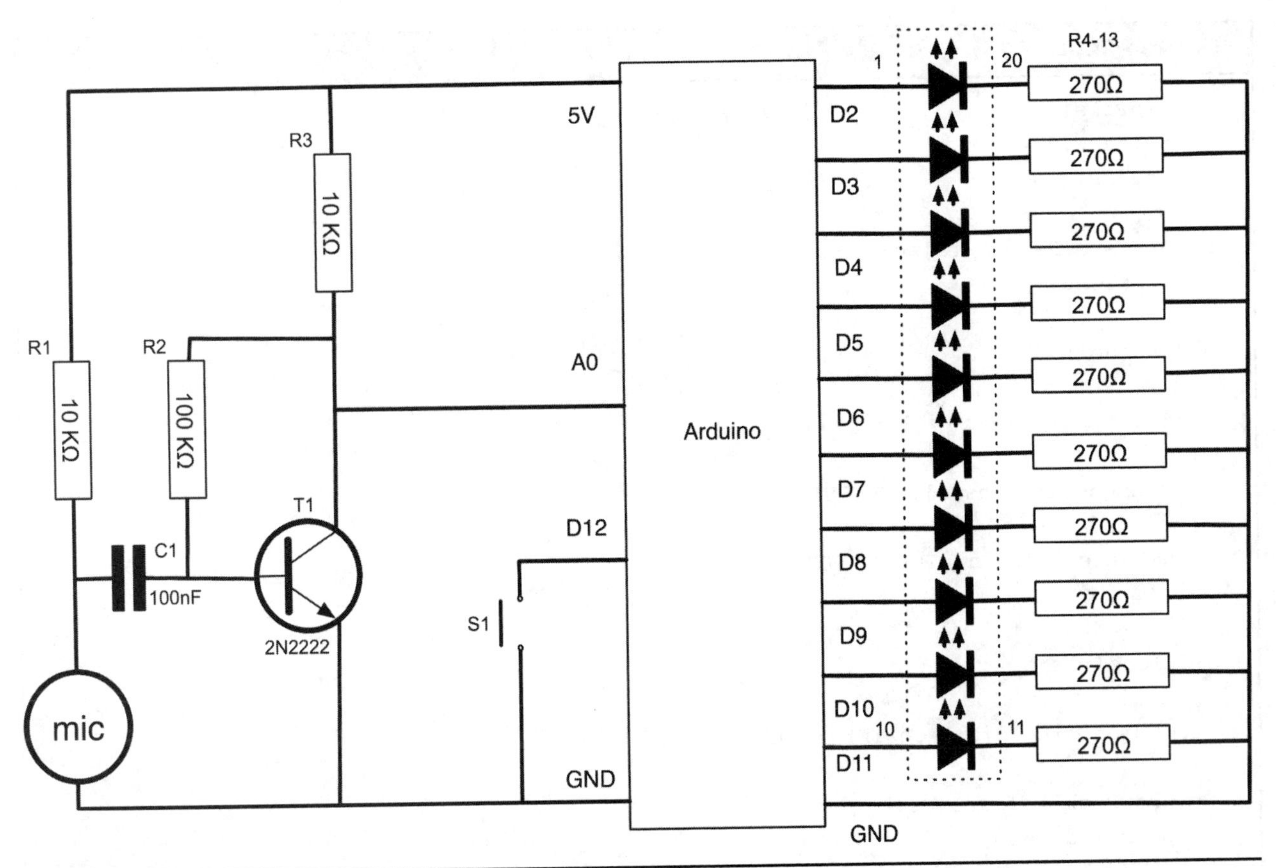

Figure 7-16 Schematic diagram for Project 21.

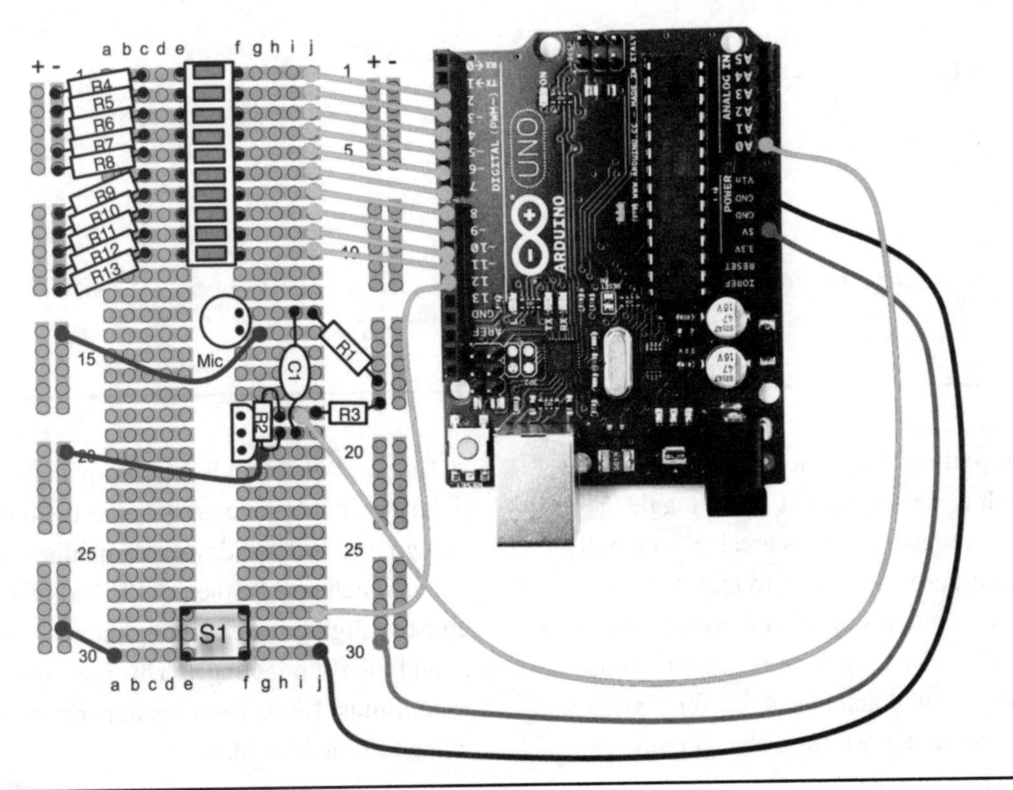

Figure 7-17 Breadboard layout for Project 21.

LISTING PROJECT 21

```
int ledPins[] = {2, 3, 4, 5, 6, 7, 8, 9, 10, 11};
int switchPin =12;
int soundPin = 0;

boolean showPeak = false;
int peakValue = 0;

void setup()
{
  for (int i = 0; i < 10; i++)
  {
    pinMode(ledPins[i], OUTPUT);
  }
  pinMode(switchPin, INPUT_PULLUP);
}

void loop()
{
  if (digitalRead(switchPin) == LOW)
  {
    showPeak = ! showPeak;
    peakValue = 0;
    delay(200); // debounce switch
  }
  int value = analogRead(soundPin);
  int topLED = map(value, 0, 1023, 0, 11) - 1;
  if (topLED > peakValue)
  {
    peakValue = topLED;
  }
  for (int i = 0; i < 10; i++)
  {
      digitalWrite(ledPins[i], (i <= topLED || (showPeak && i == peakValue)));
  }
}
```

At the top of the loop function, we check to see if the switch is depressed; if it is, we toggle the mode. The ! command inverts a value, so it will turn true into false and false into true. For this reason, it is sometimes referred to as the *marketing operator*. After changing the mode, we reset the maximum value to 0 and then delay for 200 ms to prevent keyboard bounce from changing the mode straight back again.

The level of sound is read from analog pin 0, and then we use the map function to convert from a range of 0 to 1023 down to a number between 0 and 9, which will be the top LED to be lit. This is adjusted slightly by extending the range up to 0 to 11 and then subtracting 1. This prevents the two bottom-most LEDs from being permanently lit owing to transistor bias.

We then iterate over the numbers 0 to 9 and use a Boolean expression that returns true (and hence lights the LED) if i is less than or equal to the top LED. It is actually more complicated than this because we also should display that LED if we are in peak mode and that LED happens to be the peakValue.

Putting It All Together

Load the completed sketch for Project 21 from your Arduino Sketchbook and download it to the board (see Chapter 1).

Summary

This concludes our sound-based projects. In Chapter 8 we go on to look at how we use an Arduino board to control power—a topic always close to the heart of the Evil Genius.

CHAPTER 8

Power Projects

HAVING LOOKED AT LIGHT and sound, the Evil Genius now turns his or her attention to controlling power. In essence, this means turning things on and off and controlling their speed. This mostly applies to motors and lasers and the long-awaited servo-controlled laser project.

Project 22
LCD Thermostat

The temperature in the Evil Genius' lair must be regulated because the Evil Genius is particularly susceptible to chills. This project uses an LCD screen and a temperature sensor to display both the current temperature and the set temperature. It uses a rotary encoder to allow the set temperature to be changed. The rotary encoder's button also acts as an override switch.

When the measured temperature is less than the set temperature, a relay is activated. Relays are old-fashioned electromagnetic components that activate a mechanical switch when a current flows through a coil of wire. They have a number of advantages. First, they can switch high currents and voltages, making them suitable for controlling electrical service equipment. They also electrically isolate the control side (the coil) from the switching side so that the high and low voltages never meet, which is definitely a good thing.

If the reader decides to use this project to switch electrical service electricity, he or she should do so only if he or she really knows what to do and exercises extreme caution. Electrical service electricity is very dangerous and kills about 500 people a year in the United States alone. Many more suffer painful and damaging burns.

COMPONENTS AND EQUIPMENT

	Description	Appendix
	Arduino Uno or Leonardo	m1/m2
IC1	TMP36 temperature sensor	34
R1	270 Ω, 0.25 W resistor	r3
R2	1 kΩ, 0.25 W metal film resistor	r5
R3	10 kΩ trimpot	r11
D1	5-mm red LED	s1
D2	1N4004 diode	s12
T1	2N2222 transistor	s14
	5V relay	h16
	LCD module HD44780	m6
	Header pin strip	h12
2x	Solderless breadboard	h1
	Jumper wires	h2

Hardware

The LCD module is connected up in exactly the same way as Project 17. The rotary encoder is also connected up in the same way as previous projects.

The relay will require about 70 mA, which is a bit too much for an Arduino output to handle unaided, so we use an NPN transistor to increase the current. You will also notice that a diode is connected in parallel with the relay coil. This is to prevent something called *back EMF* (electromotive force), which occurs when the relay is turned off. The sudden collapse of the magnetic field in the coil generates a voltage that can be high enough to damage the electronics if the diode is not there to effectively short it out if it occurs.

Figure 8-1 shows the schematic diagram for the project.

This project actually requires two half-sized breadboards or one single full-size breadboard. Even with two breadboards, the breadboard layout for the project is quite cramped because the LCD module uses a lot of the space.

Check your datasheet for the relay because the connection pins can be quite counterintuitive and there are several pin layouts, and your layout may not be the same as the relay that the author used.

Figure 8-2 shows the breadboard layout for the project.

You can also use a multimeter to find the coil connections by putting it on resistance mode. There will be only a pair of pins with a resistance of 40 to 100 Ω.

Software

The software for this project borrows heavily from several of our previous projects: the LCD display, the temperature data logger, and the traffic signal

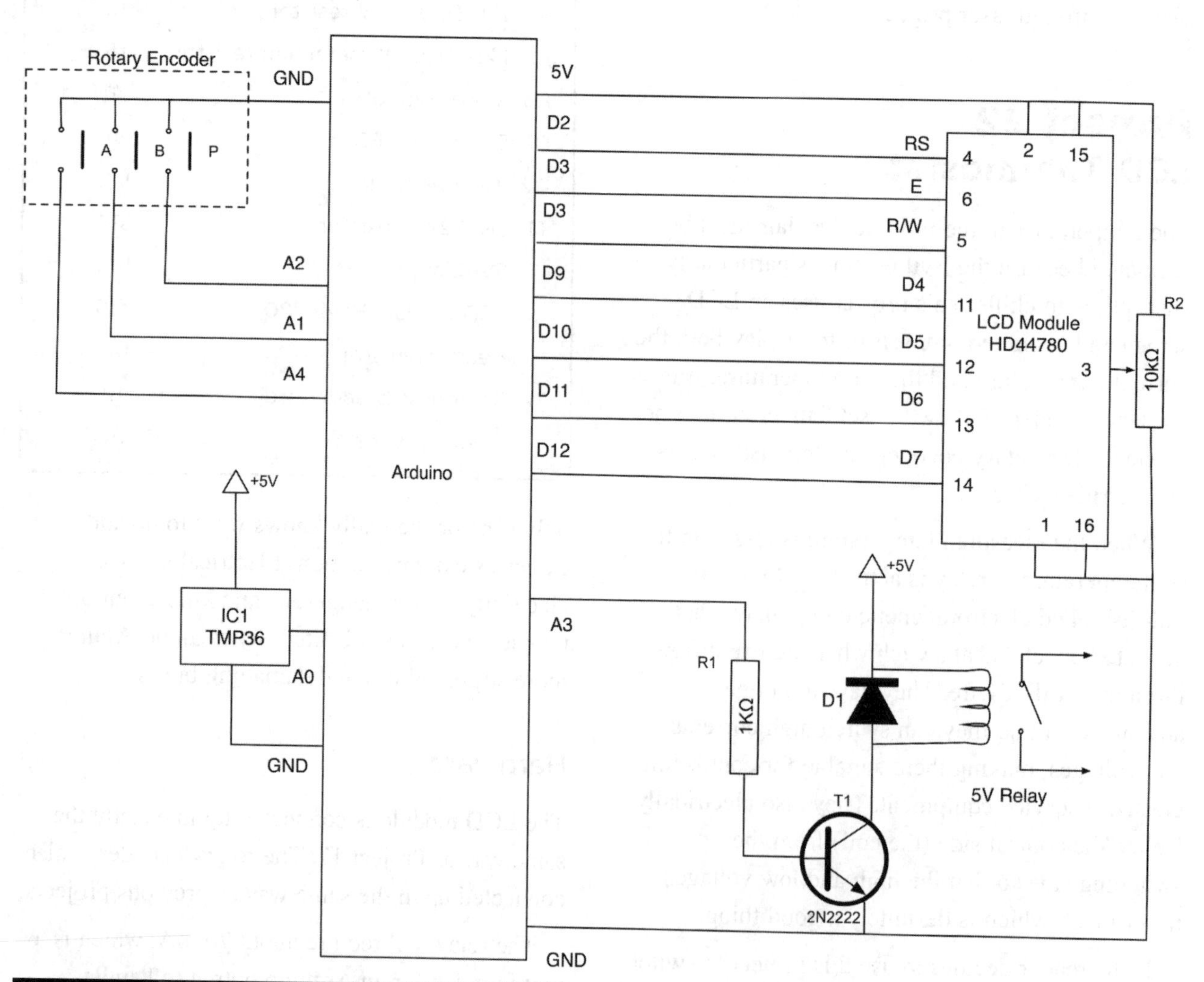

Figure 8-1 Schematic diagram for Project 22.

project for use of the rotary encoder (see Listing Project 22).

One thing that requires a bit of consideration when designing a thermostat like this is that you want to avoid what is called *hunting*. Hunting occurs when you have a simple on-off control system. When the temperature falls below the set point, the power is turned on, and the room heats until it is above the set point. Then the room cools until the temperature is below the set point again, at which point the heat is turned on again, and so on. This may take a little time to happen, but when the temperature is just balanced at the switchover temperature, this hunting can be frequent. High-frequency switching such as this is undesirable because turning things on and off tends to wear them out. This is true of relays as well.

One way to minimize this effect is to introduce something called *hysteresis*, and you may have noticed a variable called hysteresis in the sketch that is set to a value of 0.25°C.

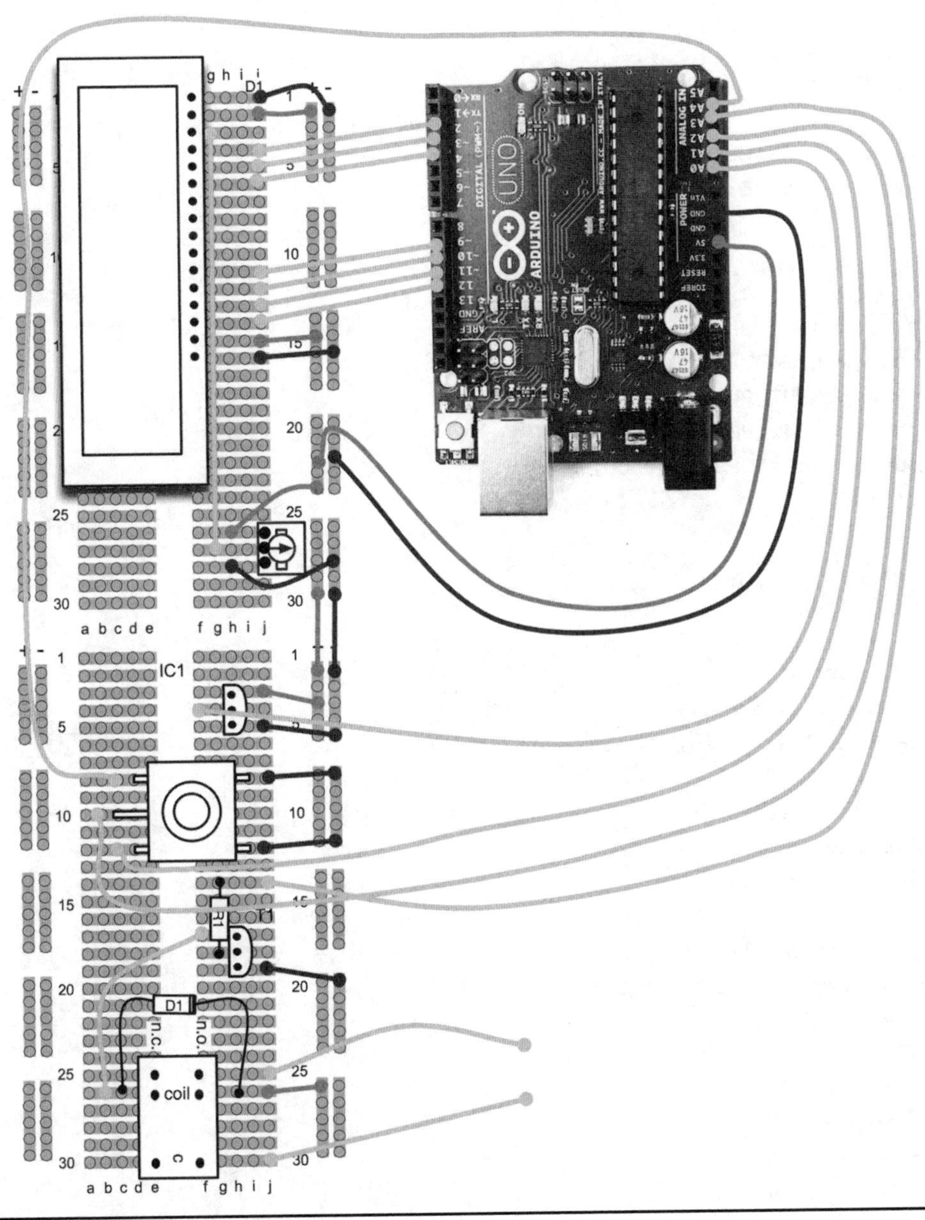

Figure 8-2 Breadboard layout for Project 22.

LISTING PROJECT 22

```
// Project 22 - LCD Thermostat

#include <LiquidCrystal.h>

// LiquidCrystal(rs, rw, enable, d4, d5, d6, d7)
LiquidCrystal lcd(2, 3, 4, 9, 10, 11, 12);

int relayPin = A3;
int aPin = A4;
int bPin = A1;
int buttonPin = A2;
int analogPin = A0;

float setTemp = 20.0;
float measuredTemp;
char mode = 'C';                // can be changed to F
boolean override = false;
float hysteresis = 0.25;

void setup()
{
  lcd.begin(2, 16);
  pinMode(relayPin, OUTPUT);
  pinMode(aPin, INPUT_PULLUP);
  pinMode(bPin, INPUT_PULLUP);
  pinMode(buttonPin, INPUT_PULLUP);
  lcd.clear();
}

void loop()
{
  static int count = 0;
  measuredTemp = readTemp();
  if (digitalRead(buttonPin) == LOW)
  {
    override = ! override;
    updateDisplay();
    delay(500); // debounce
  }
  int change = getEncoderTurn();
  setTemp = setTemp + change * 0.1;
  if (count == 1000)
  {
    updateDisplay();
    updateOutput();
    count = 0;
  }
```

LISTING PROJECT 22 (*continued*)

```
  count ++;
}

int getEncoderTurn()
{
  // return -1, 0, or +1
  static int oldA = LOW;
  static int oldB = LOW;
  int result = 0;
  int newA = digitalRead(aPin);
  int newB = digitalRead(bPin);
  if (newA != oldA || newB != oldB)
  {
    // something has changed
    if (oldA == LOW && newA == HIGH)
    {
      result = -(oldB * 2 - 1);
    }
  }
  oldA = newA;
  oldB = newB;
  return result;
}

float readTemp()
{
  int a = analogRead(analogPin);
  float volts = a / 205.0;
  float temp = (volts - 0.5) * 100;
  return temp;
}

void updateOutput()
{
  if (override || measuredTemp < setTemp - hysteresis)
  {
    digitalWrite(relayPin, HIGH);
  }
  else if (!override && measuredTemp > setTemp + hysteresis)
  {
    digitalWrite(relayPin, LOW);
  }
}

void updateDisplay()
{
```

(continued on next page)

LISTING PROJECT 22 (*continued*)

```
    lcd.setCursor(0,0);
    lcd.print("Actual: ");
    lcd.print(adjustUnits(measuredTemp));
    lcd.print(" o");
    lcd.print(mode);
    lcd.print(" ");

    lcd.setCursor(0,1);
    if (override)
    {
      lcd.print(" OVERRIDE ON ");
    }
    else
    {
      lcd.print("Set:   ");
      lcd.print(adjustUnits(setTemp));
      lcd.print(" o");
      lcd.print(mode);
      lcd.print(" ");
    }
}

float adjustUnits(float temp)
{
   if (mode == 'C')
   {
     return temp;
   }
   else
   {
     return (temp * 9) / 5 + 32;
   }
}
```

Figure 8-3 shows how we use a hysteresis value to prevent high-frequency hunting.

As the temperature rises with the power on, it approaches the set point. However, it does not turn off the power until it has exceeded the set point plus the hysteresis value. Similarly, as the temperature falls, the power is not reapplied the moment it falls below the set point but only when it falls below the set point minus the hysteresis value.

We do not want to update the display continuously because any tiny changes in the reading would result in the display flickering wildly. So, instead of updating the display every time round the main loop, we just do it one time in 1000. This still means that it will update three or four times per second. To do this, we use the technique of having a counter variable that we increment each time round the loop. When it gets to 1000, we update the display and reset the counter to 0.

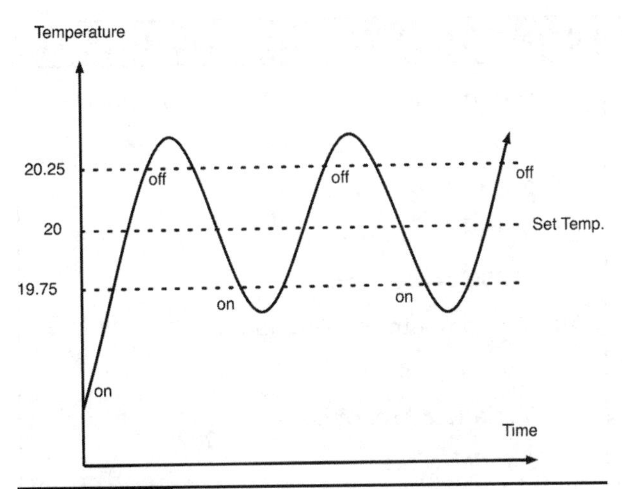

Figure 8-3 Hysteresis in control systems.

Using lcd.clear() each time we change the display also would cause it to flicker. So we simply write the new temperatures on top of the old temperature. This is why we pad the OVERRIDE ON message with spaces so that any text that was previously displayed at the edges will be blanked out.

Putting It All Together

Load the completed sketch for Project 22 from your Arduino Sketchbook and download it to the board (see Chapter 1).

The completed project is shown in Figure 8-4. To test the project, turn the rotary encoder, setting the set temperature to slightly above the actual temperature. The relay should click on. Then put your finger onto the TMP36 to warm it up. If all is well, then when the set temperature is exceeded, the LED should turn off, and you will hear the relay click.

You also can test the operation of the relay by connecting a multimeter in continuity test (beep) mode to the switched output leads.

Figure 8-4 Project 22: LCD thermostat.

I cannot stress enough that if you intend to use your relay to switch electrical service electricity, first put this project onto a properly soldered Protoshield because breadboard is not suitable for high voltages. Second, be very careful, and check and double-check what you are doing. Electrical service electricity kills.

You *must* only test the relay with low voltage unless you are going to make a proper soldered project from this design.

Project 23
Computer-Controlled Fan

One handy part to reclaim from a dead PC is the case fan (Figure 8-5). We are going to use one of these fans to keep ourselves cool in the summer. Obviously, a simple on-off switch would not be in keeping with the Evil Genius' way of doing things, so the speed of the fan will be controllable from our computer.

If you do not happen to have a dead computer lying around, fear not, because you can buy new cooling fans quite cheaply.

COMPONENTS AND EQUIPMENT

	Description	Appendix
	Arduino Uno or Leonardo	m1/m2
R1	270 Ω, 0.25 W resistor	r3
D1	1N4004 Diode	s12
T1	BD139 power transistor	s17
M1	12V computer cooling fan	h17
	12V power supply	h7
	Solderless breadboard	h1
	Jumper wires	h2

Hardware

We can control the speed of the fan using the analog output (pulse-width modulation) driving a power transistor to send pulses of power to the motor. Since these computer fans are usually 12V, we will use an external power supply to provide the drive power for the fan. The fan is likely to have a positive and a negative lead. The positive lead is often red.

Figure 8-6 shows the schematic diagram for the project and Figure 8-7 the breadboard layout.

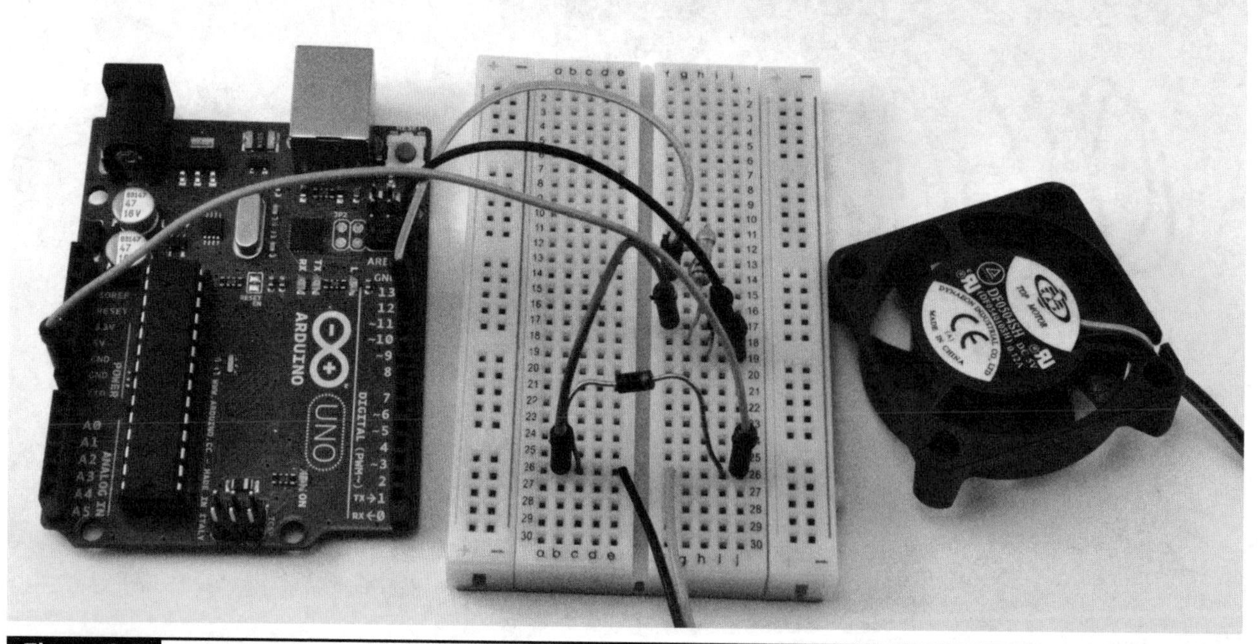

Figure 8-5 Project 23: computer-controlled fan.

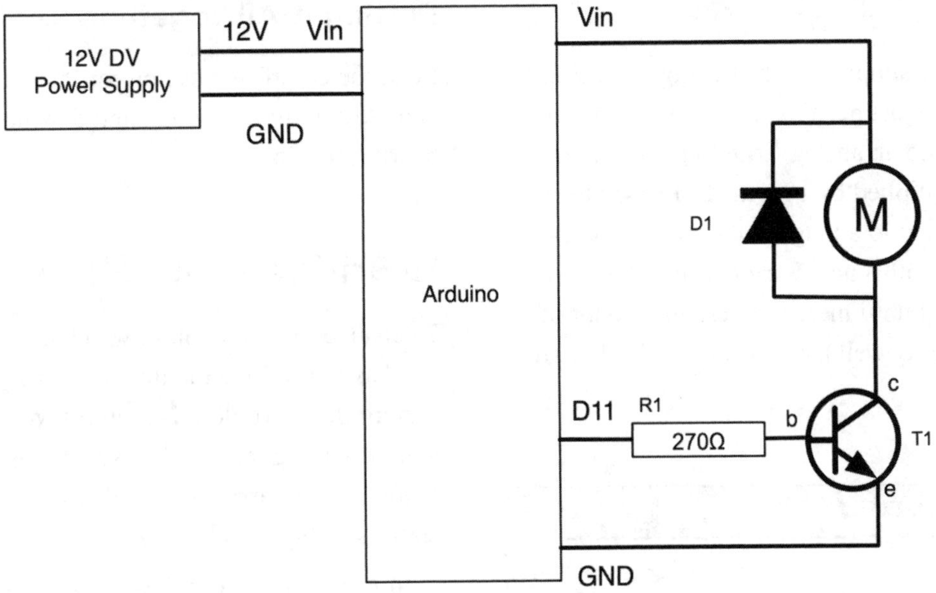

12V DV
Power Supply
12V
Vin
GND
Arduino
Vin
D1
M
D11
R1
270Ω
b
c
e
T1
GND

Figure 8-6 Schematic diagram for Project 23.

IC1
R1
T1
D1
a b c d e
f g h i j
+
12V
Fan

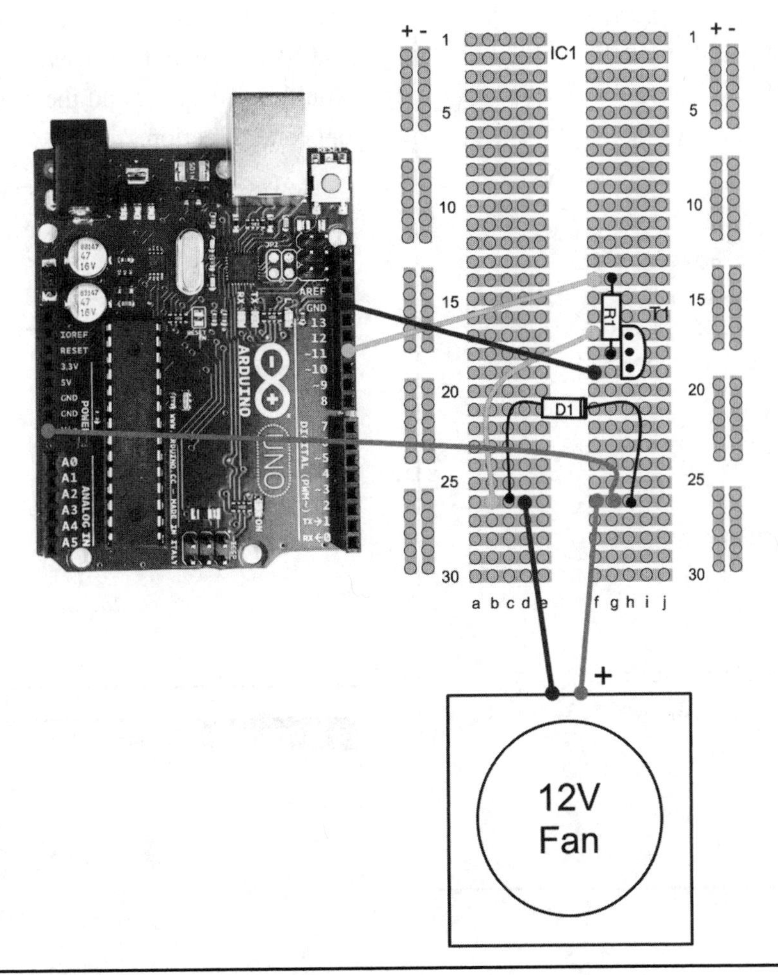

Figure 8-7 Breadboard layout for Project 23.

Software

This is a really simple sketch (Listing Project 23). Essentially, we just need to read a digit 0 to 9 from the USB and do an analogWrite to the motorPin of that value multiplied by 10 with 150 added to it. This will scale it to a range between 150 and 240. The offset of 150 is necessary because the fan will not move at all until there is a certain amount of voltage. You may well have to tweek this value for your fan.

LISTING PROJECT 23

```
// Project 23 - Computer Controlled Fan
int motorPin = 11;

void setup()
{
  pinMode(motorPin, OUTPUT);
  analogWrite(motorPin, 0);
  Serial.begin(9600);
}

void loop()
{
  if (Serial.available())
  {
    char ch = Serial.read();
    if (ch >= '0' && ch <= '9')
    {
      int speed = ch - '0';
      if (speed == 0)
      {
        analogWrite(motorPin, 0);
      }
      else
      {
        analogWrite(motorPin, 150 +
                      speed * 10);
      }
    }
  }
}
```

Putting It All Together

Load the completed sketch for Project 23 from your Arduino Sketchbook and download it to the board (see Chapter 1).

H-Bridge Controllers

To change the direction in which a motor turns, you have to reverse the direction in which the current flows. To do this requires four switches or transistors. Figure 8-8 shows how this works, using switches in an arrangement that is, for obvious reasons, called an *H-bridge.*

In Figure 8-8, S1 and S4 are closed, and S2 and S3 are open. This allows current to flow through the motor, with terminal A being positive and terminal B being negative. If we were to reverse the switches so that S2 and S3 are closed and S1 and S4 are open, then B would be positive and A would be negative, and the motor would turn in the opposite direction.

However, you may have spotted a danger with this circuit. That is, if by some chance S1 and S2

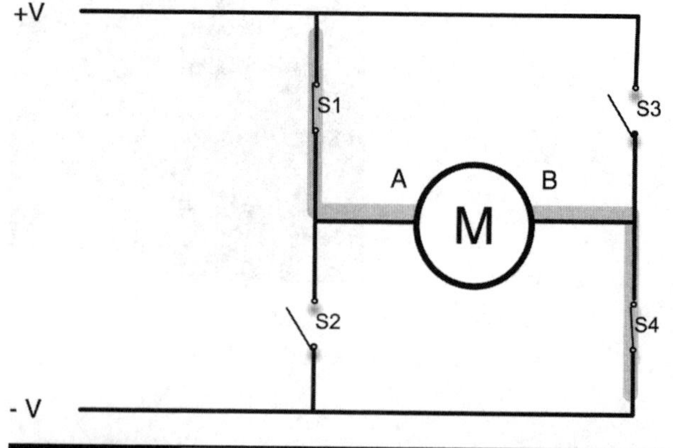

Figure 8-8 An H-bridge.

are both closed, then the positive supply will be connected directly to the negative supply, and we will have a short-circuit. The same is true if S3 and S4 are both closed at the same time.

Although you can use individual transistors to make an H-bridge, it is simpler to use an H-bridge integrated circuit (IC) such as the L293D. This chip actually has two H-bridges in it, so you can use it to control two motors. We will use one of these chips in Project 24.

Project 24
Hypnotizer

Mind control is one of the Evil Genius' favorite things. This project (see Figure 8-9) takes complete control of a motor not only to control its speed but also to make it turn clockwise and counterclockwise. Attached to the motor will be a swirling spiral disk intended to mesmerize unfortunate victims.

COMPONENTS AND EQUIPMENT

	Description	Appendix
	Arduino Uno or Leonardo	m1/m2
M1	6V DC gearmotor	h18
	Wheel	h19
IC1	L293D motor driver	s24
	Solderless breadboard	h1
	Jumper wires	h2

The motor that we use in this project is a gear motor; that is, it is a DC motor and gearbox combined into a single unit. The gearing makes the shaft turn more slowly, making it more suitable for this project.

Hardware

The schematic diagram for the hypnotizer is shown in Figure 8-10. It uses just one of the two channels available on the L293D chip.

Figure 8-9 Project 24: the hypnotizer.

The L293D has two +V pins (8 and 16). The pin +Vmotor (8) provides the power for the motors, and +V (16) supplies the chip's logic. We have connected both of these to the Arduino 5V pin. However, if you were using a more powerful motor or a higher-voltage motor, you would provide the motor with a separate power supply using pin 8 of the L293D connected to the positive power supply and the ground of the second power supply connected to the ground of the Arduino.

Figure 8-11 shows the breadboard layout for the project.

Our hypnotizer needs a spiral pattern to work. You may decide to photocopy Figure 8-12, cut it out, and stick it to the fan. Alternatively, a more colorful version of the spiral is available to print out from www.arduinoevilgenius.com.

The spiral was cut out of paper and stuck onto cardboard that was then glued onto the little cog on the end of the motor.

Software

The sketch uses an array, speeds, to control the disk's progression in speed. This makes the disk spin faster and faster in one direction and then slow until it eventually reverses direction and then starts getting faster and faster in that direction, and so on. You may need to adjust this array for your particular motor. The speeds you will need to specify in the array will vary from motor to motor, so you probably will need to adjust these values.

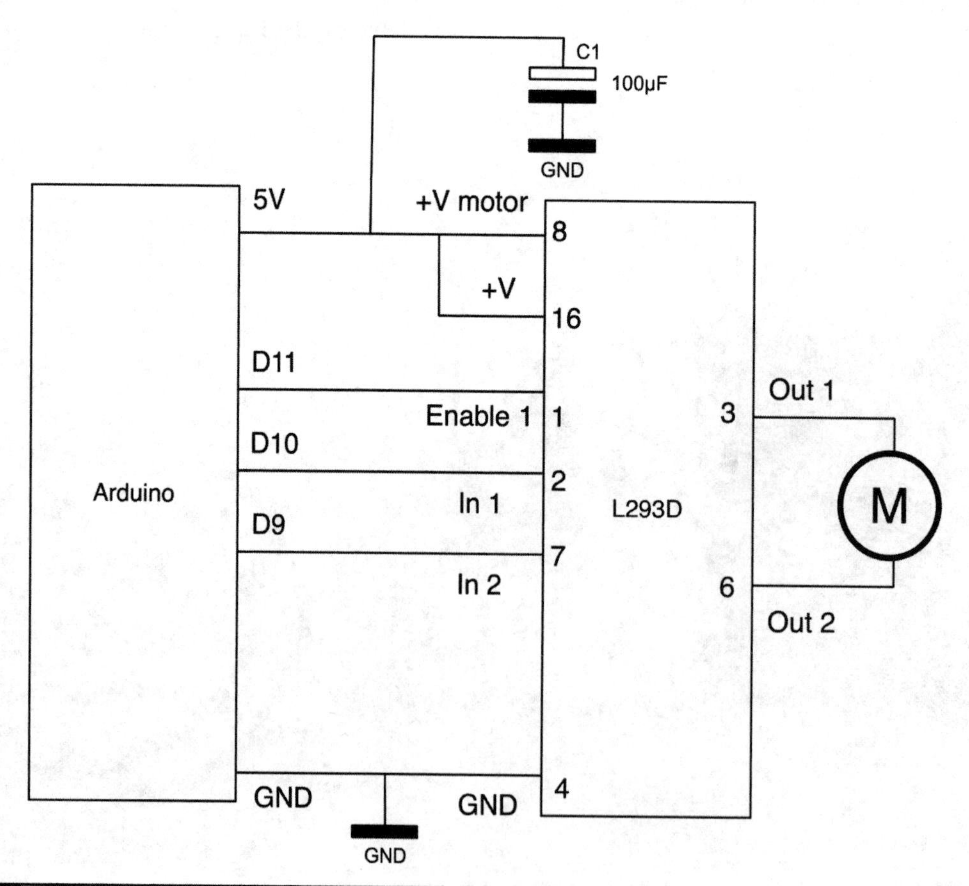

Figure 8-10 Schematic diagram for Project 24.

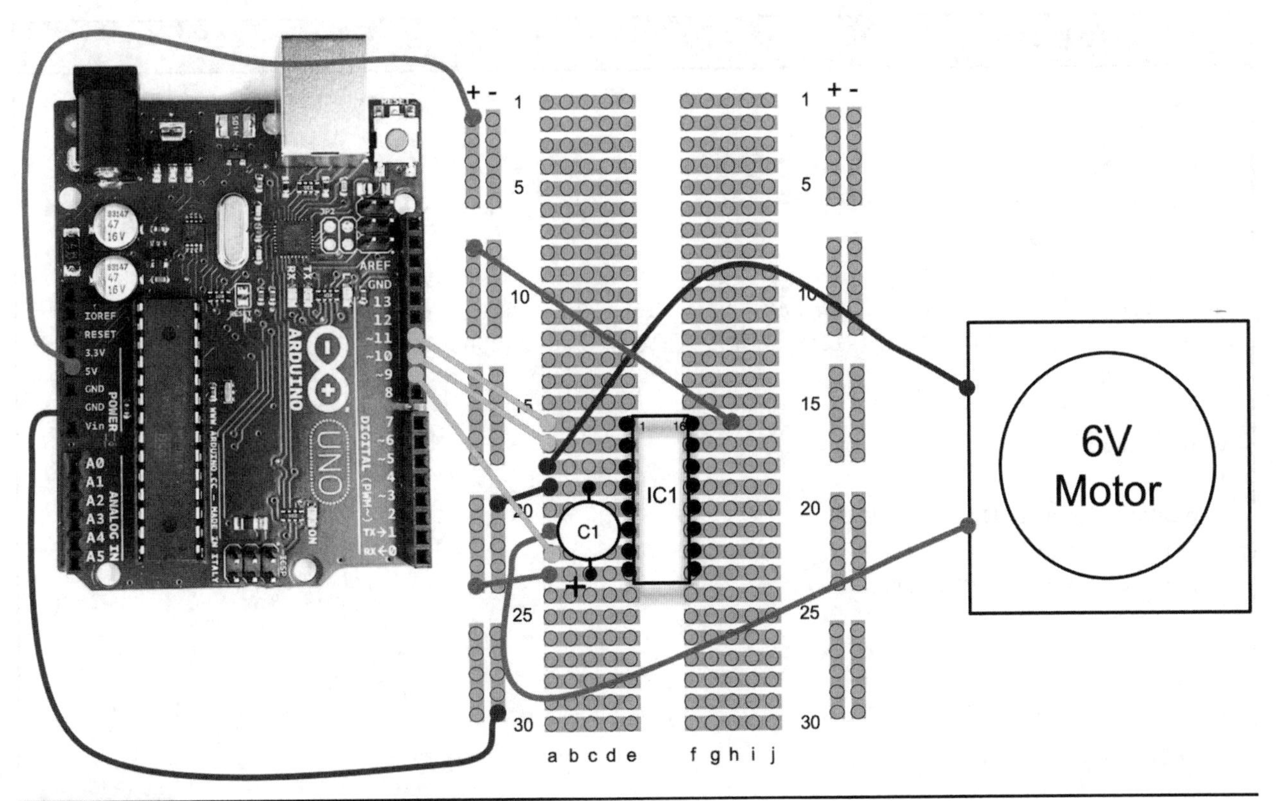

Figure 8-11 Breadboard layout for Project 24.

The enable pin of the chip controls the speed of the motor using PWM, and the in1 and in2 pins control the direction of the motor as shown in the following table:

In1	In2	Motor
GND	GND	Stopped
5V	GND	Turns in direction A
GND	5V	Turns in direction B
5V	5V	Stopped

Figure 8-12 Spiral for the hypnotizer.

Putting It All Together

Load the completed sketch for Project 24 from your Arduino Sketchbook and download it to the board (see Chapter 1).

Take care to check your wiring before applying power on this project. You can test each path through the H-bridge by connecting the control wires that go to digital pins 5 and 6 to ground.

LISTING PROJECT 24

```
// Project 24 - Hypnotizer

int enable1Pin = 11;
int in1Pin = 10;
int in2Pin = 9;

int speeds[] = {80, 100, 160, 240, 250, 255, 250, 240, 160, 100, 80,
               -80, -100, -160, -240, -250, -255, -250, -240, -160, -100, -80};
int i = 0;

void setup()
{
  pinMode(enable1Pin, OUTPUT);
  pinMode(in1Pin, OUTPUT);
  pinMode(in2Pin, OUTPUT);
}

void loop()
{
  int speed = speeds[i];
  i++;
  if (i == 22)
  {
    i = 0;
  }
  drive(speed);
  delay(1500);
}

void drive(int speed)
{
  if (speed > 0)
  {
    analogWrite(enable1Pin, speed);
    digitalWrite(in1Pin, HIGH);
    digitalWrite(in2Pin, LOW);
  }
  else if (speed < 0)
  {
    analogWrite(enable1Pin, -speed);
    digitalWrite(in1Pin, LOW);
    digitalWrite(in2Pin, HIGH);
  }
```

Then connect one of the leads to 5V, and the motor should turn in one direction. Connect that lead back to ground, and then connect the other lead to 5V, and the motor should rotate in the other direction.

Servo Motors

Servo motors are great little components that are often used in radio-controlled cars to control steering and in model aircraft to move the control surfaces. Servo motors come in a variety of sizes for different types of applications, and their wide use in models makes them relatively inexpensive.

Unlike normal motors, they do not rotate continuously; rather, you set them to a particular angle using a PWM signal. They contain their own control electronics to do this, so all you have to provide them with is power (which, for many devices, can be 5V) and a control signal that you can generate from the Arduino board.

Over the years, the interface to servos has become standardized. The servo must receive a continuous stream of pulses at least every 20 ms. The angle that the servo maintains is determined by the pulse width. A pulse width of 1.5 ms will set the servo at its midpoint, or 90 degrees. A pulse of 1.75 ms normally will swing it round to 180 degrees, and a shorter pulse of 1.25 ms will set the angle to 0 degrees.

Project 25
Servo-Controlled Laser

This project (see Figure 8-13) uses two servo motors to aim a laser diode. It can move the laser quite quickly so that you can "write" on distant walls using it.

Figure 8-13 Project 25: servo-controlled laser.

COMPONENTS AND EQUIPMENT

	Description	Appendix
	Arduino Uno or Leonardo	m1/m2
D1	3 mW red laser module	s11
M1, M2	9g servo motor	h20
R1	100 Ω, 0.25W resistor	r2
C1	100 QF capacitor	r3
	Solderless breadboard	h1
	Jumper wires	h2

This is a real laser. It is not high-powered, only 3 mW, but nonetheless, do not shine the beam in your own or anybody else's eyes. To do so could cause retina damage.

Hardware

The schematic diagram for the project is shown in Figure 8-14. It is all quite simple. The servos have just three leads. For each servo, the brown lead is connected to ground, the red lead to +5V, and the orange (control) lead to digital outputs 2 and 3.

The servos are terminated in sockets designed to fit over a pin header. Jumper wires are used to connect these to the breadboard.

The laser module is driven just like an ordinary LED from D4 via a current-limiting resistor.

The servos are usually supplied with a range of "arms" that push onto a cogged drive and are secured by a retaining screw. One of the servos is glued onto one of these arms (Figure 8-15). Then the arm is attached to the servo. Do not fit the retaining screw yet because you will need to adjust the angle. Glue the laser diode to a second arm, and attach that to the servo. It is a good idea to fix some of the wire from the laser to the arm to prevent strain on the wire where it emerges from the laser. You can do this by putting a loop of solid-core wire through two holes in the server arm and twisting it around the lead. You can see this in Figure 8-17.

You now need to attach the bottom servo to the breadboard. Self-adhesive putty will hold it in place firmly enough. Make sure that you understand how the servo will move before you

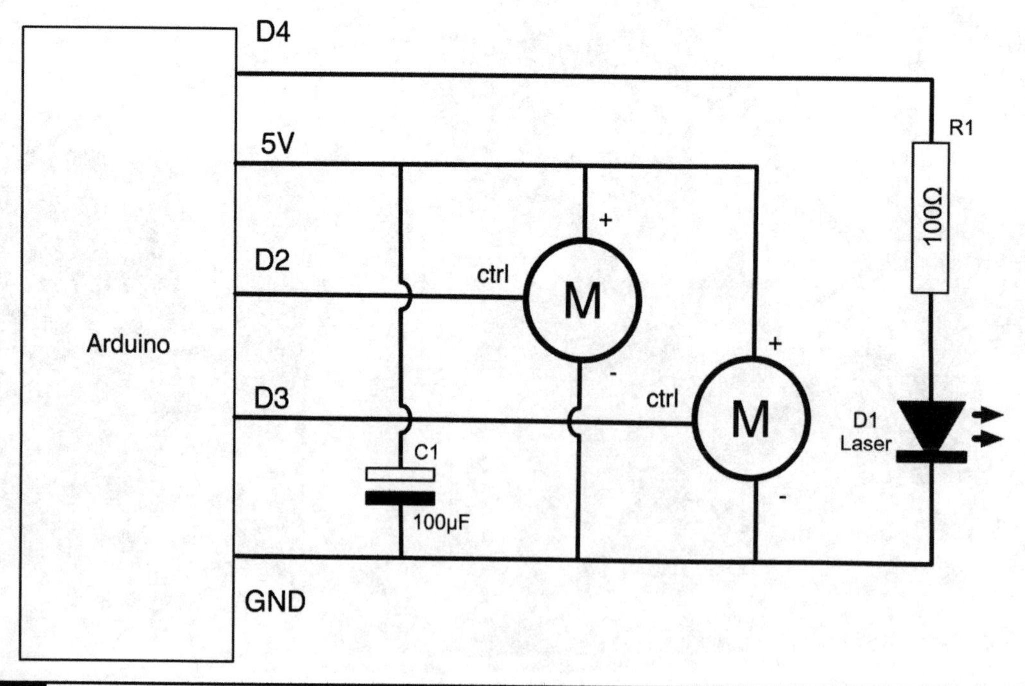

Figure 8-14 Schematic diagram for Project 25.

Figure 8-15 Attaching one servo to an arm.

glue the bottom servo to anything. If in doubt, wait until you have installed the software, and try the project out just holding the bottom servo before you glue it in place. Once you are sure that everything is in the right place, fit the retaining screws onto the servo arms.

You can see how the breadboard is used to anchor the various wires in Figure 8-16. There are no components except the resistor and capacitor on the breadboard.

Different servos draw different amounts of current. If you find that your Arduino resets itself whenever the servos move, then powering the Arduino from an external 9V or 12V adapter should fix this.

Software

Fortunately for us, a servo library comes with the Arduino library, so all we need to do is tell each servo what angle to set itself at. There is obviously more to it than this because we want to have a means of issuing our evil project with coordinates at which to aim the laser.

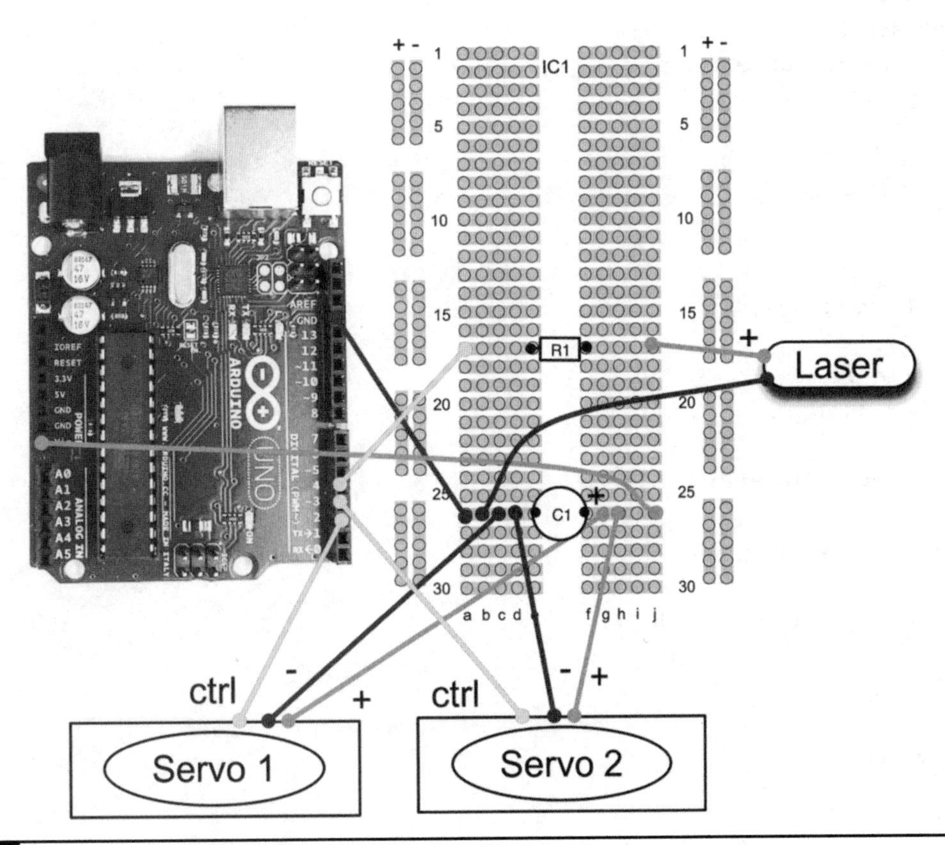

Figure 8-16 Breadboard layout for Project 25.

To do this, we allow commands to be sent over a USB cable. The commands are in the form of letters. R, L, U, and D direct the laser right, left, up, and down, respectively, by 5 degrees. For finer movements, r, l, u, and d move the laser by just 1 degree. To pause and allow the laser to finish moving, you can send the dash (–) character. (See Listing Project 25.)

LISTING PROJECT 25

```
#include <Servo.h>

int laserPin = 4;
Servo servoV;
Servo servoH;

int x = 90;
int y = 90;
int minX = 10;
int maxX = 170;
int minY = 50;
int maxY = 130;

void setup()
{
  servoH.attach(3);
  servoV.attach(2);
  pinMode(laserPin, OUTPUT);
  Serial.begin(9600);
}

void loop()
{
  char ch;
  if (Serial.available())
  {
    ch = Serial.read();
    if (ch == '0')
    {
      digitalWrite(laserPin, LOW);
    }
    else if (ch == '1')
    {
      digitalWrite(laserPin, HIGH);
    }
    else if (ch == '-')
    {
      delay(100);
    }
    else if (ch == 'c')
```

LISTING PROJECT 25 (*continued*)

```
      {
        x = 90;
        y = 90;
      }
      else if (ch == 'l' || ch == 'r' || ch == 'u' || ch == 'd')
      {
        moveLaser(ch, 1);
      }
      else if (ch == 'L' || ch == 'R' || ch == 'U' || ch == 'D')
      {
        moveLaser(ch, 5);
      }
    }
    servoH.write(x);
    servoV.write(y);
    delay(15);
}

void moveLaser(char dir, int amount)
{
  if ((dir == 'r' || dir == 'R') && x > minX)
  {
    x = x - amount;
  }
  else if ((dir == 'l' || dir == 'L') && x < maxX)
  {
    x = x + amount;
  }
  else if ((dir == 'u' || dir == 'U') && y < maxY)
  {
    y = y + amount;
  }
  else if ((dir == 'd' || dir == 'D') && x > minY)
  {
    y = y - amount;
  }
}
```

There are three other commands. The letter c will center the laser back at its resting position, and the commands 1 and 0 turn the laser on and off, respectively.

Putting It All Together

Load the completed sketch for Project 25 from your Arduino Sketchbook and download it to the board (see Chapter 1).

Open up the Serial Monitor, and type the following sequence. You should see the laser trace the letter A, as shown in Figure 8-17:

```
c1UUUUUU–RRRR–DDDDDD–0UUU–1LLLL–0DDD
```

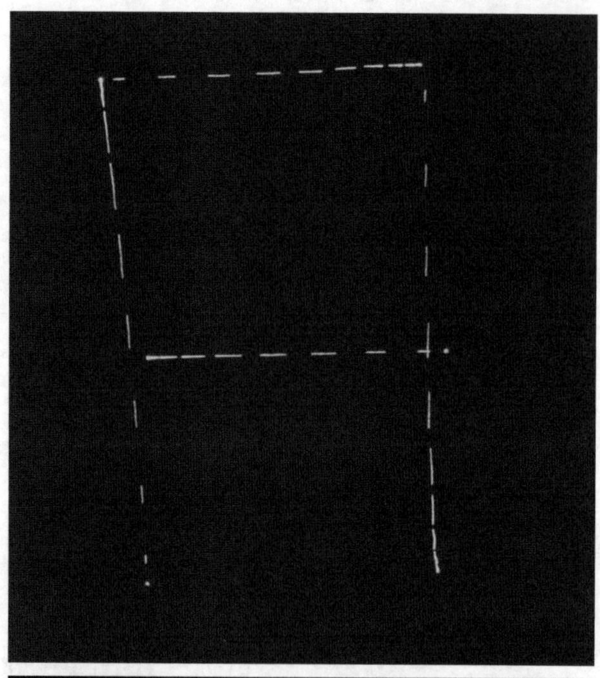

Figure 8-17 Writing the letter A with the laser.

Summary

In previous chapters we have built up our knowledge of how to use light, sound, and various sensors on the Arduino. We also have learned how to control the power to motors and to use relays. This covers nearly everything we are likely to want to do with our Arduino board, so in Chapter 9 we can put all these things together to create some wider-ranging projects.

CHAPTER 9

Miscellaneous Projects

THIS CHAPTER IS JUST A collection of projects that we can build. They do not illustrate any particular point except that Arduino projects are great fun to make.

Project 26
Lie Detector

How can an Evil Genius be sure that his or her prisoners are telling the truth? By using a lie detector, of course. This lie detector (Figure 9-1) uses an effect known as *galvanic skin response*.

As a person becomes nervous—for example, when telling a lie—his or her skin resistance decreases. We can measure this resistance using an analog input and use an LED and buzzer to indicate an untruth.

WARNING Because this project requires your test subject to touch electrodes on either side of your heart, then there is a very small risk that something could go wrong with your computer and output high voltage on the USB port. To avoid any chance of this, power the Arduino with a battery.

Figure 9-1 Project 26: lie detector.

We use a multicolor LED that will display red to indicate a lie, green to indicate a truth, and blue to show that the lie detector should be adjusted by twiddling the variable resistor.

There are two types of piezo buzzers. Some are just a piezoelectric transducer, whereas some also include an electronic oscillator to drive them. In this project we want the former, more common type without the electronics because we are going to generate the necessary frequency from the Arduino board itself.

Hardware

The subject's skin resistance is measured by using the subject as one resistor in a potential divider and a fixed resistor as the other. The lower the subject's resistance, the more analog input 0 will be pulled toward 5V. The higher the resistance, the closer to GND it will become.

The piezo buzzer, despite the level of noise these things generate, is actually quite low in current consumption and can be driven directly from an Arduino digital pin.

COMPONENTS AND EQUIPMENT

	Description	Appendix
	Arduino Uno or Leonardo	m1/m2
R1-3	270 Ω, 0.25 W resistor	r3
R4	470 kΩ, 0.25 W resistor	r9
R5	10 kΩ trimpot	r11
D1	RGB LED (common cathode)	s7
S1	Piezo buzzer	h21
	Thumbtacks	
	Solderless breadboard	h1
	Jumper wires	h2

This project uses the same multicolor LED as Project 14. In this case, however, we are not going to blend different colors but just turn one of the LEDs on at a time to display red, green, or blue.

Figure 9-2 shows the schematic diagram for the project and Figure 9-3 the breadboard layout.

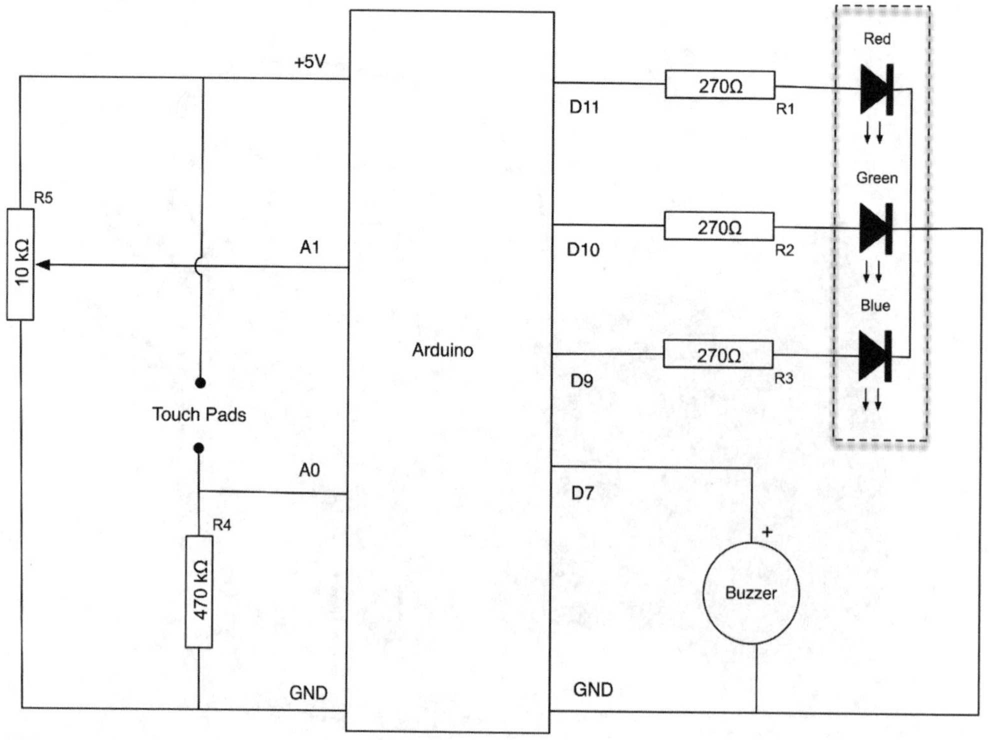

Figure 9-2 Schematic diagram for Project 26.

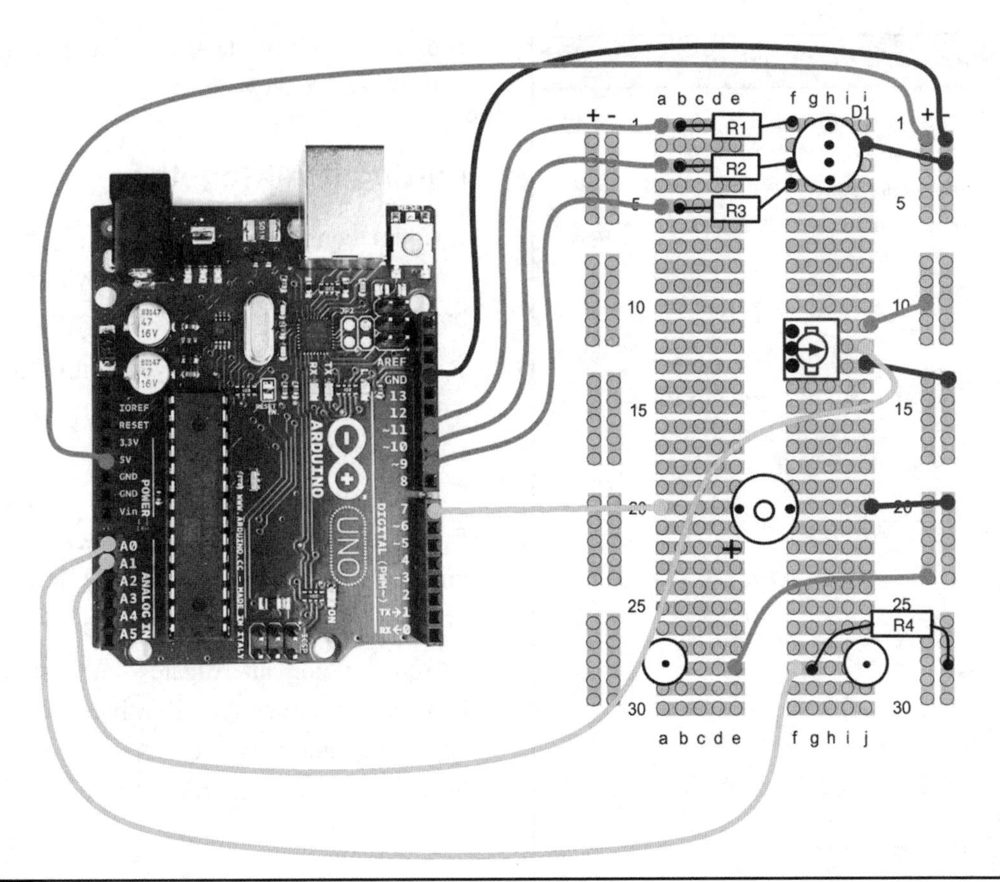

Figure 9-3 Breadboard layout for Project 26.

The variable resistor is used to adjust the set point of resistance, and the touch pads are just two metal thumbtacks pushed into the breadboard.

Software

The script for this project (Listing Project 26) just has to compare the voltage at A0 and A1. If they are about the same, the LED will be set to green. If the voltage from the finger sensor (A0) is significantly higher than A1, the variable resistor will indicate a fall in skin resistance, the LED will change to red, and the buzzer will sound. On the other hand, if A0 is significantly lower than A1, the LED will turn blue, indicating a rise in skin resistance.

The buzzer requires a frequency of about 5 kHz, or 5000 cycles per second, to drive it. We accomplish this with a simple for loop with

LISTING PROJECT 26

```
int redPin = 11; // todo paste in
                 // modified sketch
int greenPin = 10;
int bluePin = 9;
int buzzerPin = 7;

int potPin = 1;
int sensorPin = 0;

long red = 0xFF0000;
long green = 0x00FF00;
long blue = 0x000080;

int band = 10;
  // adjust for sensitivity

void setup()
{
```

(continued on next page)

LISTING PROJECT 26 (*continued*)

```
  pinMode(redPin, OUTPUT);
  pinMode(greenPin, OUTPUT);
  pinMode(bluePin, OUTPUT);
  pinMode(buzzerPin, OUTPUT);
}

void loop()
{
  int gsr = analogRead(sensorPin);
  int pot = analogRead(potPin);
  if (gsr > pot + band)
  {
    setColor(red);
    beep();
  }
  else if (gsr < pot - band)
  {
    setColor(blue);
  }
  else
  {
    setColor(green);
  }
}


void setColor(long rgb)
{
  int red = rgb >> 16;
  int green = (rgb >> 8) & 0xFF;
  int blue = rgb & 0xFF;
  analogWrite(redPin, red);
  analogWrite(greenPin, green);
  analogWrite(bluePin, blue);
}

void beep()
{
  // 5 Khz for 1/5th second
  for (int i = 0; i < 1000; i++)
  {
    digitalWrite(buzzerPin, HIGH);
    delayMicroseconds(100);
    digitalWrite(buzzerPin, LOW);
    delayMicroseconds(100);
  }
}
```

commands to turn the appropriate pin on and off with delays in between.

Putting It All Together

Load the completed sketch for Project 26 from your Arduino Sketchbook and download it to the board (see Chapter 1).

To test the lie detector, you really need a test subject because you will need one hand free to adjust the knob.

First, get your subject to place two adjoining fingers on the two metal thumbtacks. Then turn the knob on the variable resistor until the LED turns green.

You may now interrogate your victim. If the LED changes to either red or blue, you should adjust the knob until it changes to green again and then continue the interrogation.

Project 27
Magnetic Door Lock

This project (Figure 9-4) is based on Project 10 but extends it so that when the correct code is entered, it lights a green LED in addition to operating a magnetic door latch. The sketch is also improved so that the secret code can be changed without having to modify and install a new script. The secret code is stored in electrically erasable programmable read-only memory (EEPROM), so if the power is disconnected, the code will not be lost.

When powered, the electromagnetic latch will release the latch mechanism itself so that the door can be opened. When no power is applied, the latch stays in a closed position.

The DC adapter needs to be able to supply enough current to activate the latch. So check the specification for your latch before selecting a power supply. Normally, 2 A will be fine.

Figure 9-4 Project 27: magnetic door lock.

COMPONENTS AND EQUIPMENT

	Description	Appendix
	Arduino Uno or Leonardo	m1/m2
D1	5-mm red LED	s1
D2	5-mm green LED	s2
R1-2	270 Ω, 0.25 W resistor	r3
K1	4 by 3 keypad	h11
	0.1-inch header strip	h12
T1	FQP30N06 transistor	s16
	Magnetic door latch	h23
D3	1N4004	38
	Solderless breadboard	h1
	Jumper wires	h2
	12V, 2 A DC power adapter	h7

Note that these latches are designed to open only for a few seconds to allow the door to be opened.

Hardware

The schematic diagram (Figure 9-5) and breadboard layout (Figure 9-6) are much the same as for Project 10, but with additional components. Like relays, the electromagnetic latch is an inductive load and therefore liable to generate a back electromotive force (EMF), which diode D3 protects against.

The latch is controlled by T1 and switched at 12V. Because the project will be powered from a 12V adaptor, the Vin connection of the Arduino is connected to one connection of the latch.

Software

The software for this project is, as you would expect, similar to that for Project 10 (see Listing Project 27).

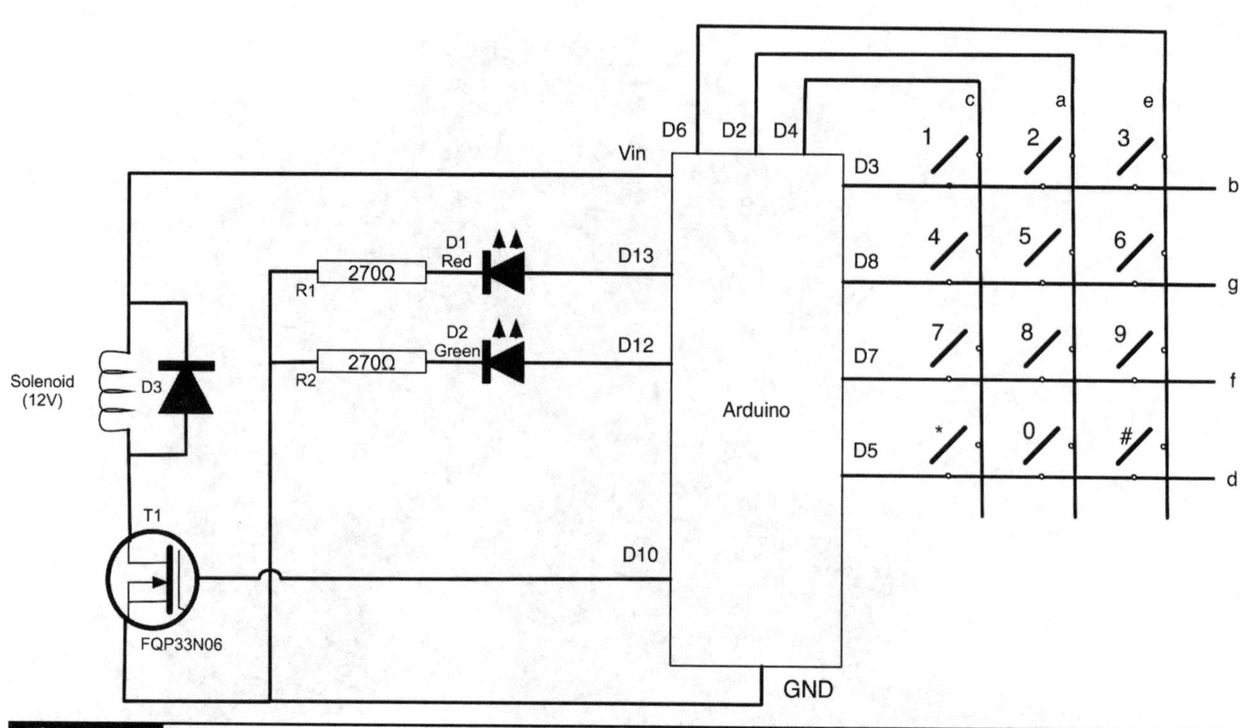

Figure 9-5 Schematic diagram for Project 27.

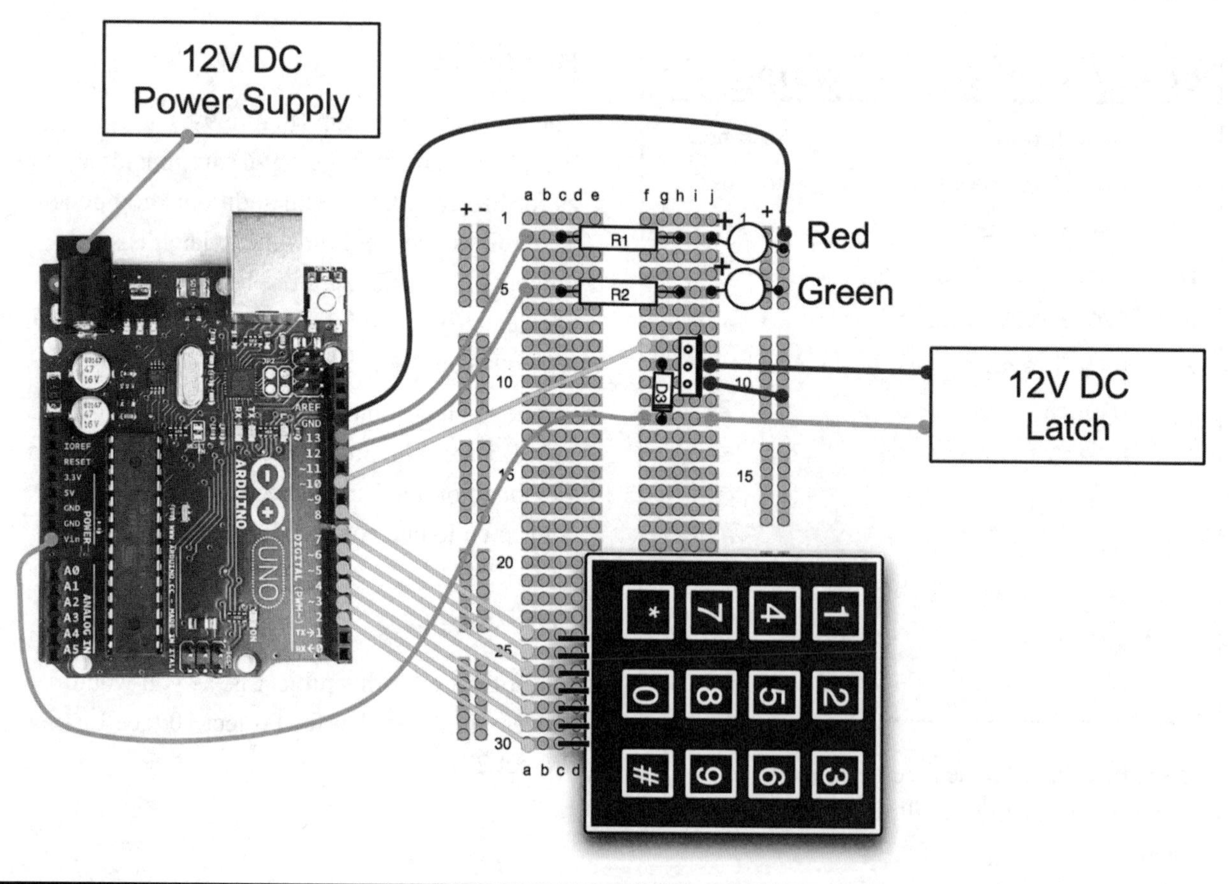

Figure 9-6 Breadboard layout for Project 27.

LISTING PROJECT 27

```
// Project 27 Keypad door lock

#include <Keypad.h>
#include <EEPROM.h>

char* secretCode = "1234";
int position = 0;

const byte rows = 4;
const byte cols = 3;
char keys[rows][cols] = {
  {'1','2','3'},
  {'4','5','6'},
  {'7','8','9'},
  {'*','0','#'}
};
byte rowPins[rows] = {7, 2, 3, 5};
byte colPins[cols] = {6, 8, 4};
Keypad keypad =
Keypad(makeKeymap(keys), rowPins,
colPins, rows, cols);

int redPin = 13;
int greenPin = 12;
int solenoidPin = 10;

void setup()
{
  pinMode(redPin, OUTPUT);
  pinMode(greenPin, OUTPUT);
  pinMode(solenoidPin, OUTPUT);
  loadCode();
  flash();
  lock();
  Serial.begin(9600);
  while(!Serial);
  Serial.print("Code is: ");
  Serial.println(secretCode);
  Serial.println("Change code: cNNNN");
  Serial.println("Unloack: u");
  Serial.println("Lock: l");
}

void loop()
{
  if (Serial.available())
  {
    char c = Serial.read();
```

LISTING PROJECT 27 *(continued)*

```
    if (c == 'u')
    {
      unlock();
    }
    if (c == 'l')
    {
      lock();
    }
    if (c == 'c')
    {
      getNewCode();
    }
  }
  char key = keypad.getKey();
  if (key == '#')
  {
    lock();
  }
  if (key == secretCode[position])
  {
    position ++;
  }
  else if (key != 0)
  {
    lock();
  }
  if (position == 4)
  {
    unlock();
  }
  delay(100);
}

void lock()
{
  position = 0;
  digitalWrite(redPin, HIGH);
  digitalWrite(greenPin, LOW);
  digitalWrite(solenoidPin, LOW);
  Serial.println("LOCKED");
}

void unlock()
{
  digitalWrite(redPin, LOW);
  digitalWrite(greenPin, HIGH);
  digitalWrite(solenoidPin, HIGH);
```

(continued on next page)

LISTING PROJECT 27 (*continued*)

```
    Serial.println("UN-LOCKED");
    delay(5000);
    lock();
}


void getNewCode()
{
  for (int i = 0; i < 4; i++ )
  {
    char ch = Serial.read();
    secretCode[i] = ch;
  }
  saveCode();
  flash();flash();
  Serial.print("Code changed to: ");
  Serial.println(secretCode);
}

void loadCode()
{
  if (EEPROM.read(0) == 1)
  {
    secretCode[0] = EEPROM.read(1);
    secretCode[1] = EEPROM.read(2);
    secretCode[2] = EEPROM.read(3);
    secretCode[3] = EEPROM.read(4);
  }
}

void saveCode()
{
  EEPROM.write(1, secretCode[0]);
  EEPROM.write(2, secretCode[1]);
  EEPROM.write(3, secretCode[2]);
  EEPROM.write(4, secretCode[3]);
  EEPROM.write(0, 1);
}

void flash()
{
    digitalWrite(redPin, HIGH);
    digitalWrite(greenPin, HIGH);
    delay(500);
    digitalWrite(redPin, LOW);
    digitalWrite(greenPin, LOW);
}
```

Although this project is powered from an external adaptor, you can still attach the USB lead to your computer and issue commands to unlock the door or change the secret code.

The setup function writes some instructions for changing the secret code using the Serial Monitor. It also shows you the current code (Figure 9-7).

The loop function has two parts. First, it looks for any incoming commands from the Serial Monitor, and then it checks for key presses.

As each key is pressed, if it matches the appropriate character in the secret code, the count variable is incremented. When the count gets to 4, the latch is unlocked.

Because each character is exactly 1 byte in length, the code can be stored directly in the EEPROM. We use the first byte of EEPROM to indicate whether the code has been set. If it has not been set, the code will default to 1234. Once the code has been set, the first EEPROM byte will be given a value of 1. If we didn't get this, then the code would become whatever happened to be in the first byte of the EEPROM.

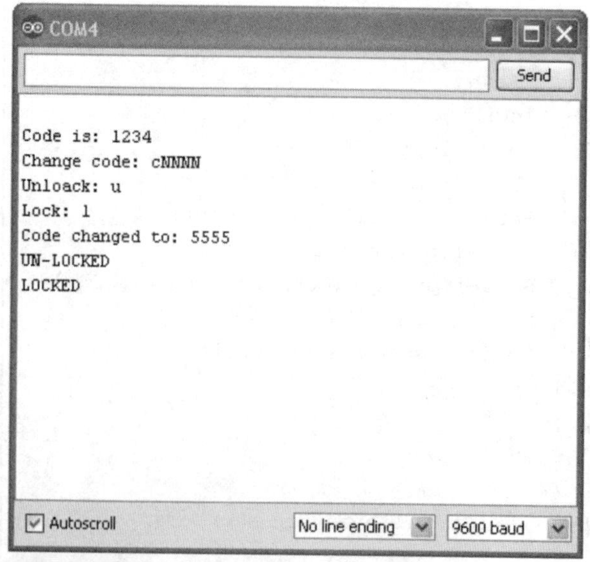

Figure 9-7 Controlling the lock with the Serial Monitor.

Putting It All Together

Load the completed sketch for Project 27 from your Arduino Sketchbook and download it to the board (see Chapter 1).

We can make sure that everything is working by powering up our project and entering the code 1234, at which point the green LED should light and the latch release.

Project 28
Infrared Remote

This project (Figure 9-8) allows the Evil Genius to control any household devices with an infrared remote control directly from his or her computer. With it, the Evil Genius can record an infrared message from an existing remote control and then play it back from his or her computer.

COMPONENTS AND EQUIPMENT

	Description	Appendix
	Arduino Uno or Leonardo	m1/m2
R1	10 Ω, 0.25 W resistor	r14
R2	270 Ω, 0.25 W resistor	r3
T1	2N2222 transistor	s14
D1	Infrared LED sender	s20
IC1	Infrared remote-control receiver	s21
	Solderless breadboard	h1
	Jumper wires	h2

We use the EEPROM to store the remote-control codes so that they are not lost when the Arduino board is disconnected.

Figure 9-8 Project 28: infrared remote.

Hardware

The infrared (IR) remote receiver is a great little module that combines an IR photodiode with all the amplification filtering and smoothing needed to produce a digital output from the IR message. This output is fed to digital pin 9. The schematic diagram (Figure 9-9) shows how simple this package is to use, with just three pins, GND, +V, and the output signal.

The IR transmitter is an IR LED. IR LEDs work just like a regular red LED but in the invisible IR end of the spectrum. On some devices you can see a slight red glow when they are on, and if you look at them using a digital camera, you can usually see the glow because digital cameras are normally slightly sensitive in the IR range.

You can power the IR sender directly from an IO pin using, say, a 270 Ω series resistor to limit the current; however, these devices are designed to be driven continuously at 100 mA (five times the current of a normal LED). So it will only have a very short range. For this reason, we are using a transistor to switch the LED and a much lower series resistor to drive the IR LED to its maximum current.

Figure 9-10 shows the breadboard layout for this project.

When building the breadboard, note that most IR LEDs defy the normal convention of LEDs. For these LEDs, the longer lead is normally the negative lead. Check with the datasheet for your LED before you wire it up.

Software

The sketch allows you to record signals from an existing remote into one of 10 memories and then play them back (see Listing Project 28).

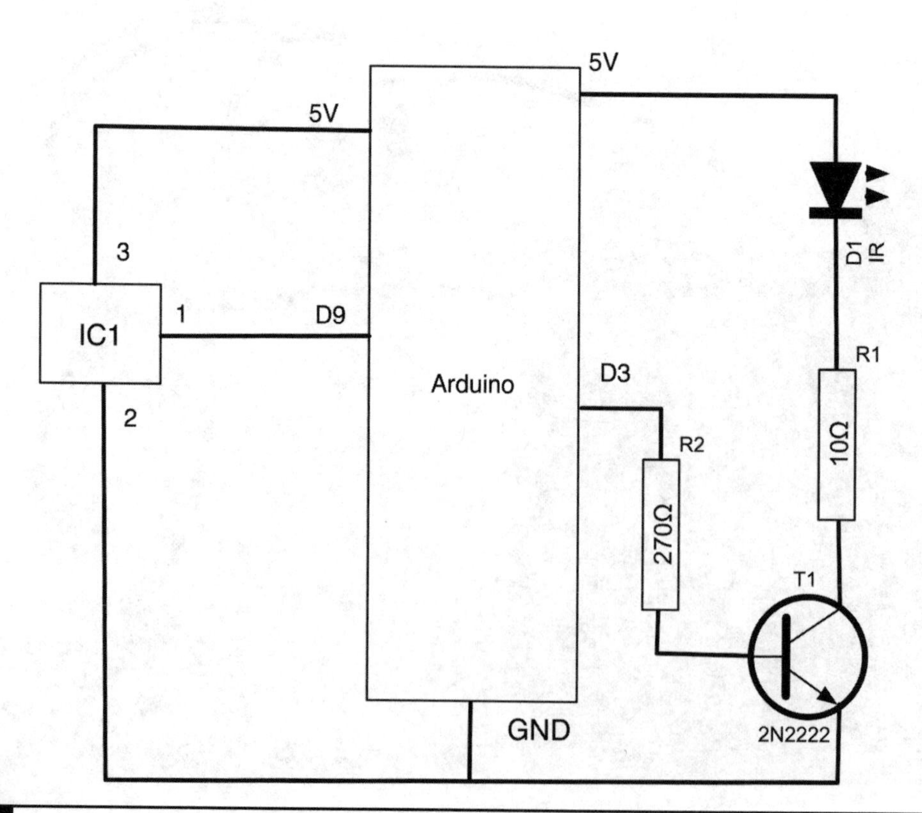

Figure 9-9 Schematic diagram for Project 28.

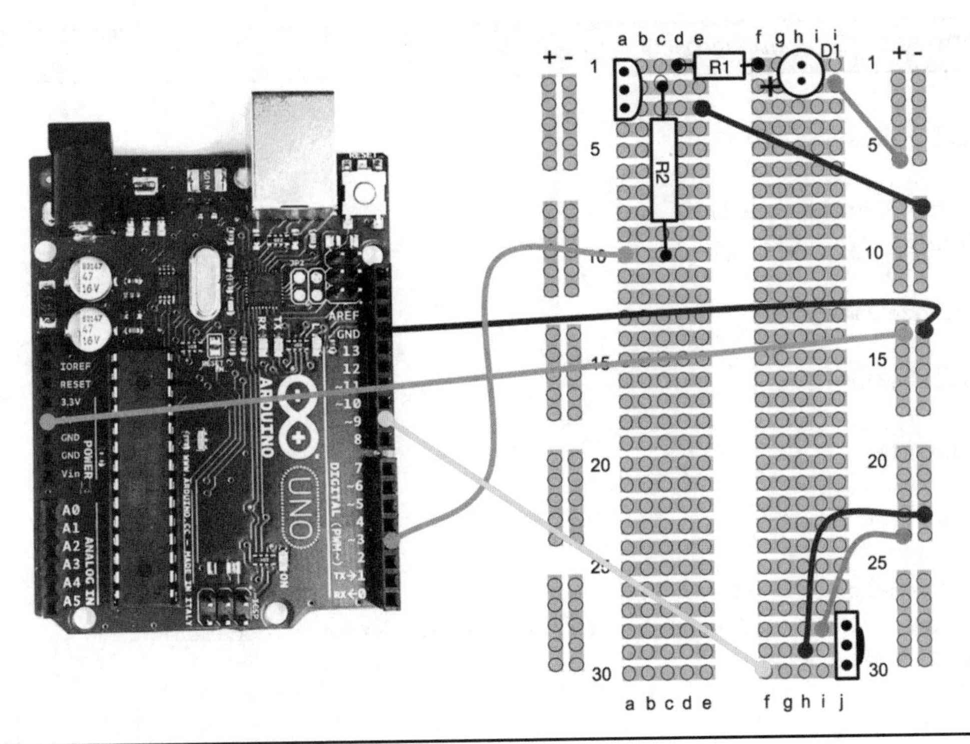

Figure 9-10 Breadboard layout for Project 28.

LISTING PROJECT 28

```
// Project 28 - IR Remote
#include <EEPROM.h>

#define maxMessageSize 100
#define numSlots 9

int irRxPin = 9;
int irTxPin = 3;

int currentCode = 0;
int buffer[maxMessageSize];

void setup()
{
  Serial.begin(9600);
  Serial.println("0-9 to set code memory, l - learn, s - to send");
  pinMode(irRxPin, INPUT);
  pinMode(irTxPin, OUTPUT);
  setCodeMemory(0);
}

void loop()
```

(continued on next page)

LISTING PROJECT 28 (*continued*)

```
{
  if (Serial.available())
  {
   char ch = Serial.read();
    if (ch >= '0' && ch <= '9')
    {
      setCodeMemory(ch - '0');
    }
    else if (ch == 's')
    {
      sendIR();
    }
    else if (ch == 'l')
    {
      int codeLen = readCode();
      Serial.print("Read code length: "); Serial.println(codeLen);
      storeCode(codeLen);
    }
  }
}

void setCodeMemory(int x)
{
  currentCode = x;
  Serial.print("Set current code memory to: ");
  Serial.println(currentCode);
}


void storeCode(int codeLen)
{
  // write the code to EEPROM, first byte is length
  int startIndex = currentCode * maxMessageSize;
  EEPROM.write(startIndex, (unsigned byte)codeLen);
  for (int i = 0; i < codeLen; i++)
  {
      EEPROM.write(startIndex + i + 1, buffer[i]);
  }
}

void sendIR()
{
  // construct a buffer from the saved data in EEPROM and send it
  int startIndex = currentCode * maxMessageSize;
  int len = EEPROM.read(startIndex);
  Serial.print("Sending Code for memory "); Serial.print(currentCode);
  Serial.print(" len="); Serial.println(len);
```

LISTING PROJECT 28 (*continued*)

```
  if (len > 0 && len < maxMessageSize)
  {
    for (int i = 0; i < len; i++)
    {
      buffer[i] = EEPROM.read(startIndex + i + 1);
    }
    sendCode(len);
  }
}

void sendCode(int n)
{
  for (int i = 0; i < 3; i++)
  {
  writeCode(n);
  delay(90);
  }
}


int readCode()
{
  int i = 0;
  unsigned long startTime;
  unsigned long endTime;
  unsigned long lowDuration = 0;
  unsigned long highDuration = 0;
  while(digitalRead(irRxPin) == HIGH) {}; // wait for first pulse
  while(highDuration < 5000l)
  {
    // find low duration
    startTime = micros();
    while(digitalRead(irRxPin) == LOW) {};
    endTime = micros();
    lowDuration = endTime - startTime;
    if (lowDuration < 5000l)
    {
      buffer[i] = (byte)(lowDuration >> 4);
      i ++;
    }
    // find the high duration
    startTime = micros();
    while(digitalRead(irRxPin) == HIGH) {};
    endTime = micros();
    highDuration = endTime - startTime;
    if (highDuration < 5000l)
```

(continued on next page)

LISTING PROJECT 28 (*continued*)

```
      {
        buffer[i] = (byte)(highDuration >> 4);
        i ++;
      }
    }
    return i;
}


void writeCode(int n)
{
    int state = 0;
    unsigned long duration = 0;
    int i = 0;
    while (i < n)
    {
        duration = buffer[i] << 4;
        int cycles = duration / 14;
        if ( ! (i % 2))
        {
          for (int x = 0; x < cycles; x++)
          {
            state = ! state;
            digitalWrite(irTxPin, state);
            delayMicroseconds(10); // less than 12 to adjust for other instructions
          }
          digitalWrite(irTxPin, LOW);
        }
        else
        {
          digitalWrite(irTxPin, LOW);
          delayMicroseconds(duration);
        }
        i ++;
    }
}
```

Infrared remote controls send a series of pulses at a frequency of between 36 and 40 kHz. Figure 9-11 shows the trace from an oscilloscope.

A bit value of 1 is represented by a pulse of square waves at 36 to 40 kHz and a 0 by a pause in which no square waves are sent.

In the setup function, we start serial communications and write instructions for using the project back to the Serial Console. It is from the Serial Console that we are going to control the project. We also set the current code memory to memory 0.

The loop function follows the familiar pattern of checking for any input through the USB port. If it is a digit between 0 and 9, it makes the corresponding memory the current memory. If an s character is received from the Serial Monitor, it

Figure 9-11 Infrared code from an oscilloscope.

sends the message in the current message memory, and if the message is l, then the sketch waits for a message to arrive from a remote.

The function then checks to see if any IR signal has been received; if it has, the function writes it to EEPROM using the storeCode function. It stores the length of the code in the first byte and then the number of 50-ms ticks for each subsequent pulse in the bytes that follow.

We also use an interesting technique in storeCode and sendIR when accessing the EEPROM that lets us use it rather like an array for the message memories. The start point for recording or reading the data from EEPROM is calculated by multiplying the currentCode by the length of each code (plus the byte that says how long it is).

Putting It All Together

Load the completed sketch for Project 28 from your Arduino Sketchbook and download it to the board (see Chapter 1).

To test the project, find yourself a remote and the bit of equipment that it controls. Then power up the project.

Open the Serial Monitor, and you should be greeted by the following message:

```
0-9 to set code memory, l - learn,
    s - to send
Set current code memory to: 0
```

By default, any message we capture will be recorded into memory 0. So enter "l" into the Serial Monitor, and then aim the remote at the sensor and press a button (turning power on and ejecting the tray on a DVD player are impressive actions). You then should see a message like this:

```
Saved code, length: 67
```

Now point the IR LED at the appliance, and type **s** into the Serial Monitor. You should receive a message like this:

```
Sent code length: 67
```

More important, the appliance should respond to the message from the Arduino board.

You now can try changing the memory slot by entering a different digit into the Serial Monitor and recording a variety of different IR commands. Note that there is no reason why they need to be for the same appliance.

Note that this project will not work on all appliances, so if it doesn't work on one, try it on another.

Project 29
Lilypad Clock

The Arduino Lilypad works in much the same way as the Uno or Leonardo boards, but instead of a boring rectangular circuit board, the Lilypad is circular and designed to be stitched into clothing using conductive thread. Even an Evil Genius appreciates beauty when he or she sees it. So this project is built into a photo frame to show off the natural beauty of the electronics (see Figure 9-12). A magnetic reed switch is used to adjust the time.

Figure 9-12 Project 29: Lilypad binary clock.

COMPONENTS AND EQUIPMENT		
	Description	**Appendix**
	Arduino Lilypad and USB programmer	m3
R1-16	100 Ω, 0.25 W resistor	r2
D1-4	2-mm red LED	s4
D5-10	2-mm blue LED	s6
D11-16	2-mm green LED	s5
R17	100 kΩ, 0.25 W metal film resistor	r8
S1	Miniature reed switch	h3
	7 x 5-inch picture frame	
	5V power supply	h6

This is a project where you have to use a soldering iron.

Hardware

We have an LED and series resistor attached to almost every connection of the Lilypad in this project.

The reed switch is a useful little component that is just a pair of switch contacts in a sealed glass envelope. When a magnet comes near the switch, the contacts are pulled together, and the switch is closed.

We use a reed switch rather than an ordinary switch so that the whole project can be mounted behind glass in a photo frame. We will be able to adjust the time by holding a magnet close to the switch.

Figure 9-13 shows the schematic diagram for the project.

Each LED has a resistor soldered to the shorter negative lead. The positive lead is then soldered to the Arduino Lilypad terminal and the lead from the resistor passes under the board, where it is connected to all the other resistor leads.

Figure 9-14 shows a close-up of the LED and resistor, and the wiring of the leads under the board is shown in Figure 9-15. Note the rough disk of paper protecting the back of the board from the soldered resistor leads.

A 5V power supply is used because a significant amount of power is used when all the LEDs are lit, so batteries would not last long. The power wires extend from the side of the picture frame, where they are soldered to a connector.

The author used a redundant cell phone power supply. Be sure to test that any supply you are going to use provides 5V at a current of at least 500 mA. You can test the polarity of the power supply using a multimeter.

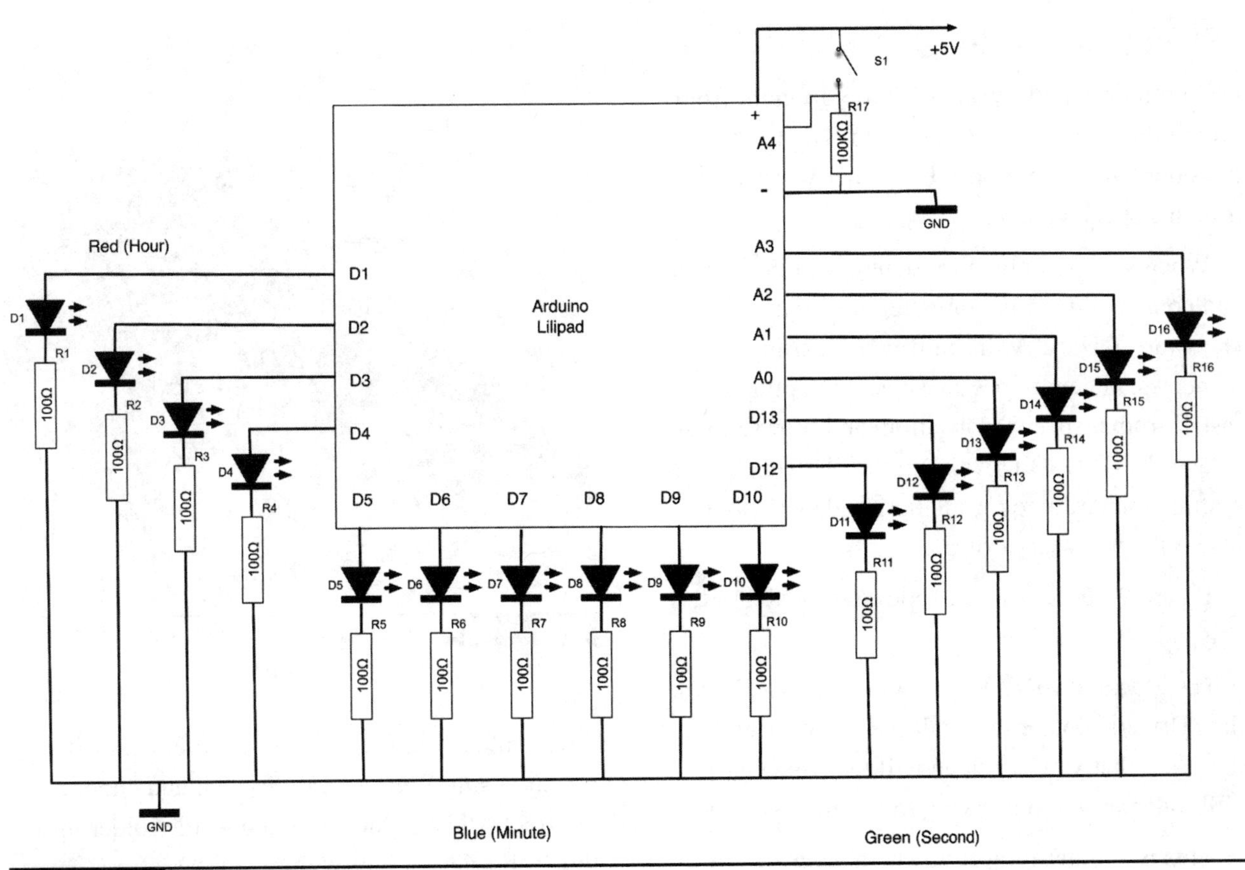

Figure 9-13 Schematic diagram for Project 29.

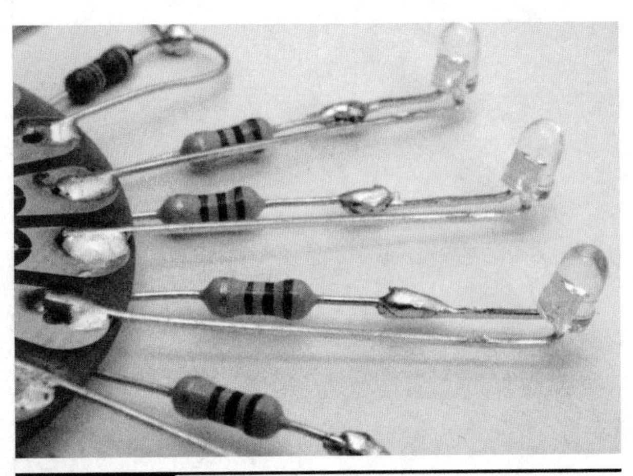

Figure 9-14 Close-up of LED attached to a resistor.

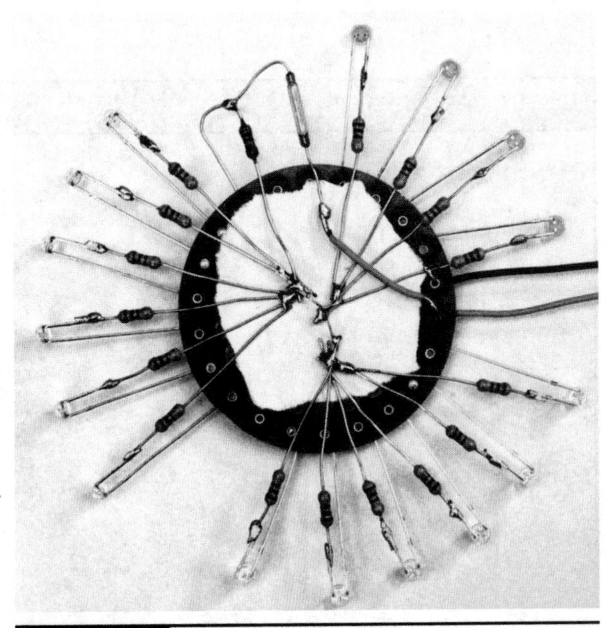

Figure 9-15 Bottom side of Lilypad board.

Software

Programming the Lilypad is a little different from programming an Uno or Leonardo. The Lilypad does not have a USB port, but rather you program it with a special adaptor.

When you first plug the adaptor into the Lilypad and connect it to your computer, the "Found New Hardware Wizard" will run if you are using Windows. When it does so, select the option to install from a specified location, and browse to the folder "FTDI USB Drivers" in the "Drivers" folder within your Arduino installation folder. This will install the necessary drivers.

Figure 9-16 shows the adaptor connected to the Lilypad.

For Mac and LINUX, you will find installers in the "Drivers" folder to install the USB driver. You may find that your machine will recognize the USB adapter without having to install anything.

This is another project in which we make use of a library. This library makes dealing with time easy and can be downloaded from http://playground.arduino.cc/Code/Time.

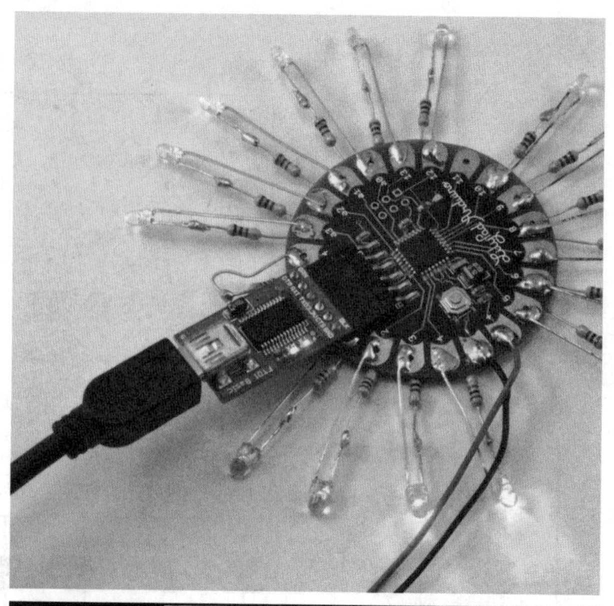

Figure 9-16 USB adaptor connected to the Lilypad board.

Download the file Time.zip, and unzip it. If you are using Windows, right-click and choose "Extract All" and then save the whole folder into the "Libraries" folder within your "Arduino sketches" folder.

Once you have installed this library into your Arduino directory, you will be able to use it with any sketches that you write (see Listing Project 29.)

LISTING PROJECT 29

```
#include <Time.h>

int hourLEDs[] = {1, 2, 3, 4};
  // least significant bit first
int minuteLEDs[] = {10, 9, 8, 7, 6, 5};
int secondLEDs[] = {17, 16, 15, 14, 13, 12};

int loopLEDs[] = {17, 16, 15, 14, 13, 12, 10, 9, 8, 7, 6, 5, 4, 3, 2, 1};

int switchPin = 18;

void setup()
{
  for (int i = 0; i < 4; i++)
  {
```

LISTING PROJECT 29 (*continued*)

```
    pinMode(hourLEDs[i], OUTPUT);
  }
  for (int i = 0; i < 6; i++)
  {
    pinMode(minuteLEDs[i], OUTPUT);
  }
  for (int i = 0; i < 6; i++)
  {
    pinMode(secondLEDs[i], OUTPUT);
  }
  setTime(0);
}

void loop()
{
  if (digitalRead(switchPin))
  {
    adjustTime(1);
  }

else if (minute() == 0 && second() == 0)
  {
    spin(hour());
  }
  updateDisplay();
  delay(1);
}

void updateDisplay()
{
  time_t t = now();
  setOutput(hourLEDs, 4, hourFormat12(t));
  setOutput(minuteLEDs, 6, minute(t));
  setOutput(secondLEDs, 6, second(t));
}


void setOutput(int *ledArray, int numLEDs, int value)
{
     for (int i = 0; i < numLEDs; i++)
     {
     digitalWrite(ledArray[i], bitRead(value, i));
     }
}
```

(continued on next page)

LISTING PROJECT 29 (*continued*)

```
void spin(int count)
{
  for (int i = 0; i < count; i++)
  {
      for (int j = 0; j < 16; j++)
      {
        digitalWrite(loopLEDs[j], HIGH);
        delay(50);
        digitalWrite(loopLEDs[j], LOW);
      }
  }
}
```

Arrays are used to refer to the different sets of LEDs. These are used to simplify installation and also in the setOutput function. This function sets the binary values of the array of LEDs that is to display a binary value. The function also receives arguments of the length of that array and the value to be written to it. This is used in the loop function to successively set the LEDs for hours, minutes, and seconds. When passing an array into a function such as this, you must prefix the argument in the function definition with an asterisk (*).

An additional feature of the clock is that every hour, on the hour, it spins the LEDs, lighting each one in turn. So at 6 o'clock, for example, it will spin six times before resuming the normal pattern.

If the reed relay is activated, the adjustTime function is called with an argument of 1 second. Because this is in the loop function with a 1-ms delay, the seconds are going to pass quickly.

Putting It All Together

Load the completed sketch for Project 29 from your Arduino Sketchbook and download it to the board. On a Lilypad, this is slightly different from what we are used to. You will have to select a different board type (Lilypad 328) and serial port from the Arduino software before downloading.

Assemble the project, but test it connected to the USB programmer before you build it into the picture frame.

Try to choose a picture frame that has a thick card insert that will allow a sufficient gap into which the components can fit between the backing board and the glass.

You may wish to design a paper insert to provide labels for your LEDs to make it easier to tell the time. A suitable design can be found at www.arduinoevilgenius.com.

To read the time from the clock, you look at each section (Hours, Minutes, and Seconds) in turn and add the values next to the LEDs that are lit. So, if the hour LEDs next to 8 and 2 are lit, then the hour is 10. Then do the same for the minutes and seconds.

Project 30
Evil Genius Countdown Timer

No book on projects for an Evil Genius should be without the Bond-style countdown timer (Figure 9-17). This timer also doubles as an egg timer because there is nothing that annoys the Evil Genius more than an overcooked soft-boiled egg!

Figure 9-17 Project 30: Evil Genius countdown timer.

COMPONENTS AND EQUIPMENT

Description	Appendix
Arduino Uno or Leonardo	m1/m2
I^2C four-digit, seven-segment display	m7
Rotary encoder	h13
Piezo buzzer	h21
Solderless breadboard	h1
Jumper wires	h2

Hardware

Like Project 16, this project also uses an I^2C module, but in this case it is a four-digit, seven-segment LED display module.

The schematic diagram for the project is shown in Figure 9-18 and the breadboard layout in Figure 9-19.

Software

The sketch for this project (Listing Project 30) uses the same libraries as Project 16. So, if you have not installed these, please refer back to Chapter 6.

Rather than make the rotary encoder change the time one second per rotation step, we have an array of standard times that fit with the egg-cooking habits of the Evil Genius. This array can be edited and extended, but if you change its length, you must alter the numTimes variable accordingly.

To keep track of the time, the function updateCountingTime checks to see if more than a second has passed, and if it has, it decrements the number of seconds by one. When the seconds get to zero, then the minute is decremented in a similar way.

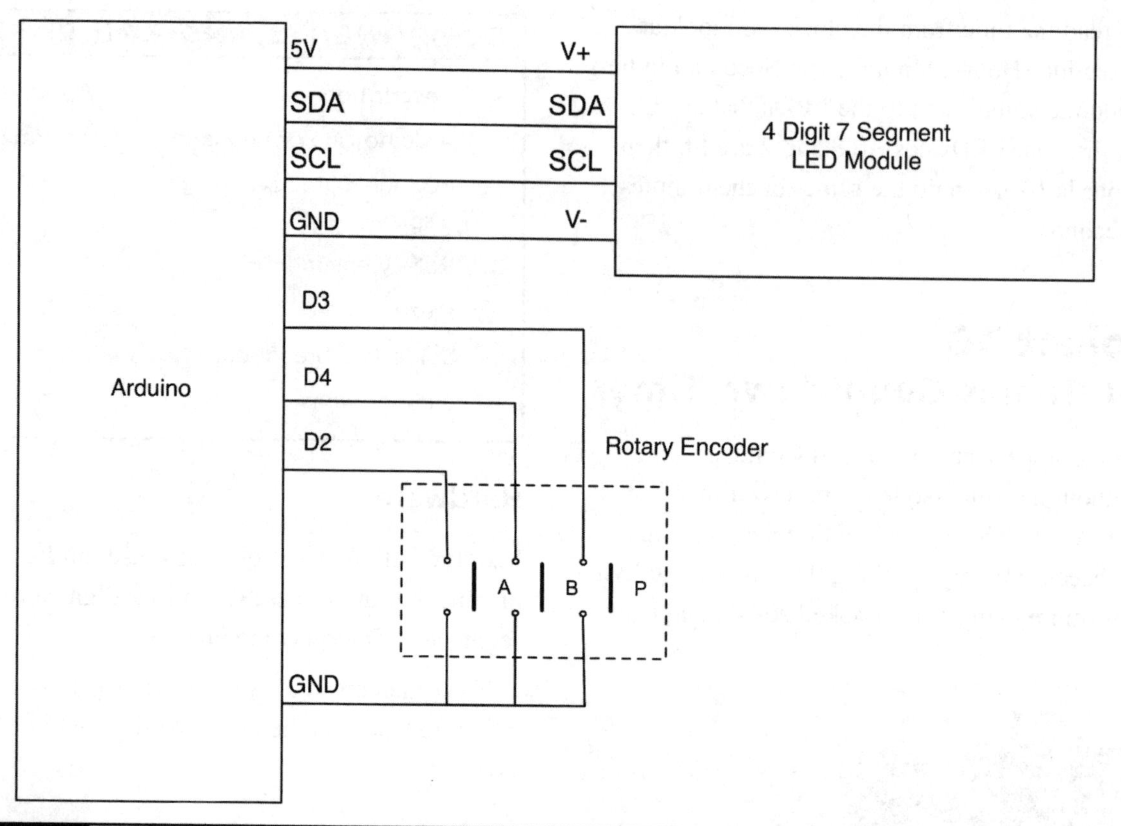

Figure 9-18 Schematic diagram for Project 30.

LISTING PROJECT 30

```
// Project 30 - Countdown Timer

#include <Adafruit_LEDBackpack.h>
#include <Adafruit_GFX.h>
#include <Wire.h>

Adafruit_7segment display = Adafruit_7segment();

int times[] = {5, 10, 15, 20, 30, 45, 100, 130, 200, 230, 300, 400, 500, 600, 700,
          800, 900, 1000, 1500, 2000, 3000};
int numTimes = 19;

int buzzerPin = 11;
int aPin = 2;
int bPin = 4;
int buttonPin = 3;

boolean stopped = true;

int selectedTimeIndex = 12;
int timerMinute;
```

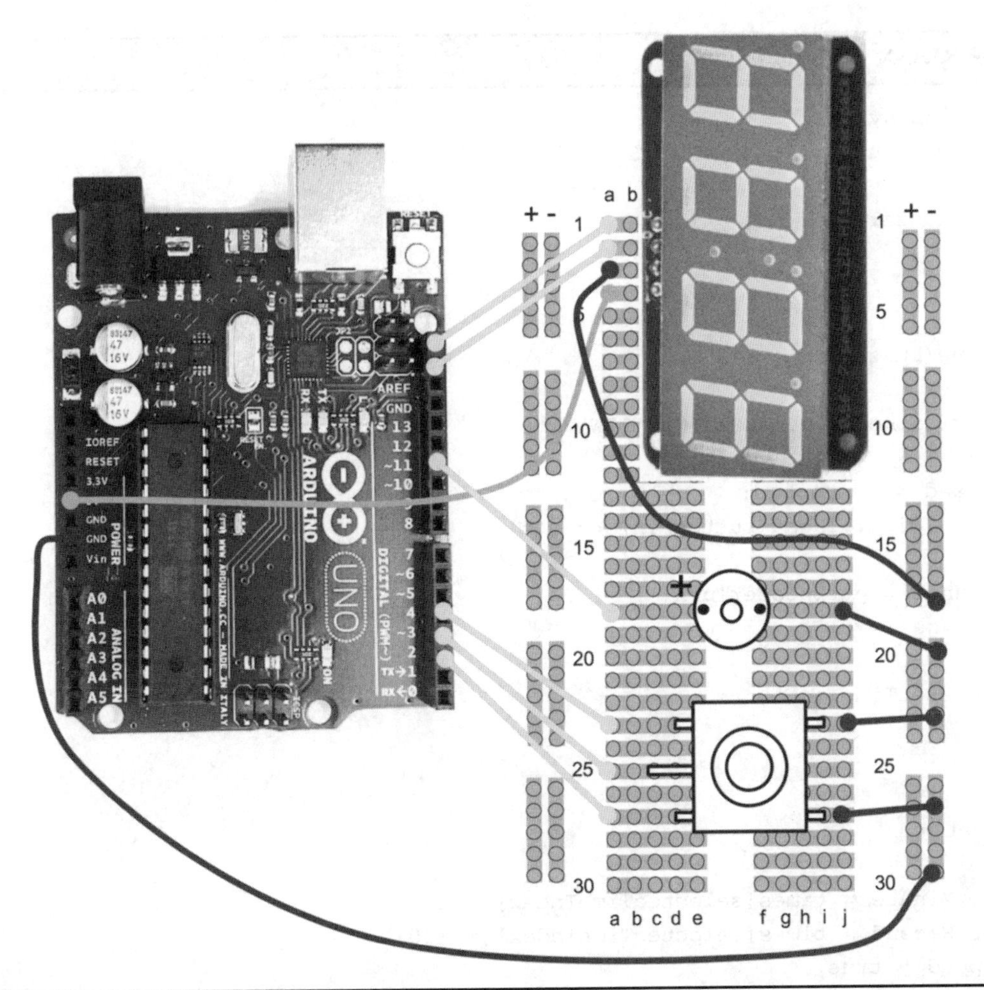

Figure 9-19 Breadboard layout for Project 30.

LISTING PROJECT 30 (*continued*)

```
int timerSecond;

void setup()
{
  pinMode(buzzerPin, OUTPUT);
  pinMode(buttonPin, INPUT_PULLUP);
  pinMode(aPin, INPUT_PULLUP);
  pinMode(bPin, INPUT_PULLUP);
  Serial.begin(9600);
  display.begin(0x70);
  reset();
}

void loop()
{
  updateCountingTime();
```

(continued on next page)

LISTING PROJECT 30 ***(continued)***

```
  updateDisplay();
  if (timerMinute == 0 && timerSecond == 0 && ! stopped)
  {
    tone(buzzerPin, 400);
  }
  else
  {
    noTone(buzzerPin);
  }
  if (digitalRead(buttonPin) == LOW)
  {
    stopped = ! stopped;
    while (digitalRead(buttonPin) == LOW);
  }
  int change = getEncoderTurn();
  if (change != 0)
  {
    changeSetTime(change);
  }
}

void reset()
{
    timerMinute = times[selectedTimeIndex] / 100;
    timerSecond = times[selectedTimeIndex] % 100;
    stopped = true;
    noTone(buzzerPin);
}

void updateDisplay() // mmss
{
  // update I2C display
  int timeRemaining = timerMinute * 100 + timerSecond;
  display.print(timeRemaining, DEC);
  display.writeDisplay();
}

void updateCountingTime()
{
  if (stopped) return;

  static unsigned long lastMillis;
  unsigned long m = millis();
  if (m > (lastMillis + 1000) && (timerSecond > 0 || timerMinute > 0))
  {
    if (timerSecond == 0)
    {
      timerSecond = 59;
```

LISTING PROJECT 30 (*continued*)

```
      timerMinute –;
    }
    else
    {
      timerSecond –;
    }
    lastMillis = m;
  }
}


void changeSetTime(int change)
{
  selectedTimeIndex += change;
  if (selectedTimeIndex < 0)
  {
    selectedTimeIndex = numTimes;
  }
  else if (selectedTimeIndex > numTimes)
  {
    selectedTimeIndex = 0;
  }
  timerMinute = times[selectedTimeIndex] / 100;
  timerSecond = times[selectedTimeIndex] % 100;
}


int getEncoderTurn()
{
  // return -1, 0, or +1
  static int oldA = LOW;
  static int oldB = LOW;
  int result = 0;
  int newA = digitalRead(aPin);
  int newB = digitalRead(bPin);
  if (newA != oldA || newB != oldB)
  {
    // something has changed
    if (oldA == LOW && newA == HIGH)
    {
      result = -(oldB * 2 - 1);
    }
  }
  oldA = newA;
  oldB = newB;
  return result;
}
```

The time to be displayed is formatted into minutes and seconds by making a single decimal number by multiplying the minutes by 100 and adding the number of seconds.

Putting It All Together

Load the completed sketch for Project 30 from your Arduino Sketchbook and download it to the board (see Chapter 1).

Summary

In Chapter 10 you will find a selection of projects designed to work with the Arduino Leonardo. This board differs from the Uno in that it can emulate a USB keyboard and mouse, opening up all sorts of possibilities.

CHAPTER 10

USB Projects with the Leonardo

The Arduino Leonardo differs from the more conventional Arduino in a number of ways. It is a little cheaper and has a different microcontroller chip. It is the use of this chip that allows the Leonardo to impersonate a USB keyboard, which is the basis for the projects described in this chapter.

Project 31
Keyboard Prank

If you are familiar with the 1999 movie, *The Matrix*—and what Evil Genius isn't?—you will remember the scene where the hero, Neo, is in his room, and messages start to appear on his computer screen.

This project uses an Arduino Leonardo secretly attached to the USB port of someone's computer to start sending those messages after a random delay (Figure 10-1).

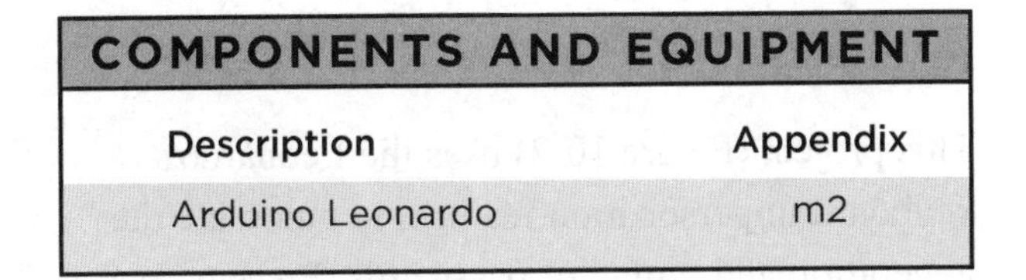

COMPONENTS AND EQUIPMENT	
Description	**Appendix**
Arduino Leonardo	m2

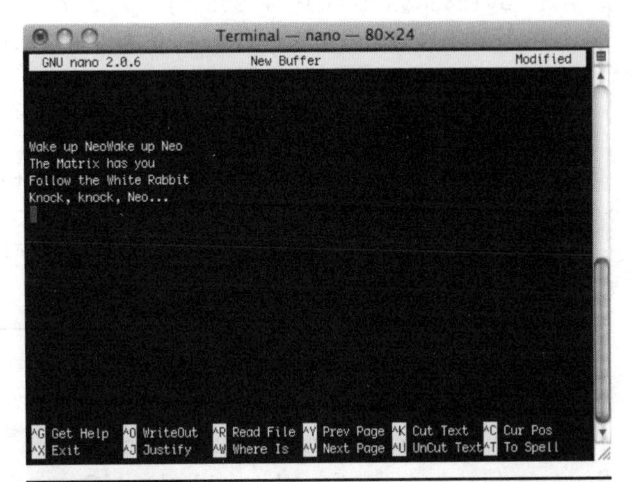

Figure 10-1 The keyboard prank in action.

Hardware

The only thing that you need for this project is a Leonardo and a USB cable, first to connect it to your computer to program it and then to connect to the computer of the person you want to prank.

Software

The sketch for Project 31 is shown in Listing Project 31.

LISTING PROJECT 31

```
// Project 31 - Keyboard Prank

void setup()
{
  randomSeed(analogRead(0));
  Keyboard.begin();
}

void loop()
{
  delay(random(10000) + 30000);
  Keyboard.print("\n\n\nWake up NeoWake up Neo\n");
  delay(random(3000) + 3000);
  Keyboard.print("The Matrix has you\n");
  delay(random(3000) + 3000);
  Keyboard.print("Follow the White Rabbit\n");
  delay(random(3000) + 3000);
  Keyboard.print("Knock, knock, Neo...\n");
}
```

The setup function seeds the random-number generator with a value from the analog input A0. Because this pin is floating, the result should be fairly random. The keyboard emulation library is also started with the command Keyboard.begin.

The main loop then waits for a random period between 30 and 40 seconds and starts sending the messages with a sorter delay of between 3 and 6 seconds between each sentence.

The \n characters are newline characters that are equivalent to pressing the ENTER key.

Because this project simulates a keyboard and starts typing all by itself, be aware that it will type the text wherever it is, and that includes the Arduino sketch if you have it open in the IDE. It is a good idea to unplug it except when you are programming the board or ready to deploy it as a prank.

If you get stuck trying to program it, a good trick is to hold down the red Reset button on the Leonardo until the Arduino software says "Uploading" in the status area, and then let go of the button.

Putting It All Together

This is a fun little project, and obviously, you can change the message text to anything you like. However, remember that the text will appear only if your victim is actually editing something where keyboard strokes will show up.

Project 32
Automatic Password Typer

This project (Figure 10-2) uses the Leonardo's keyboard impersonation features to automate the generation and typing of passwords. Pressing one key creates a new password and stores it in electrically erasable programmable read-only memory (EEPROM) so that it is not forgotten, and pressing the other key types the password using the Leonardo's ability to impersonate a keyboard.

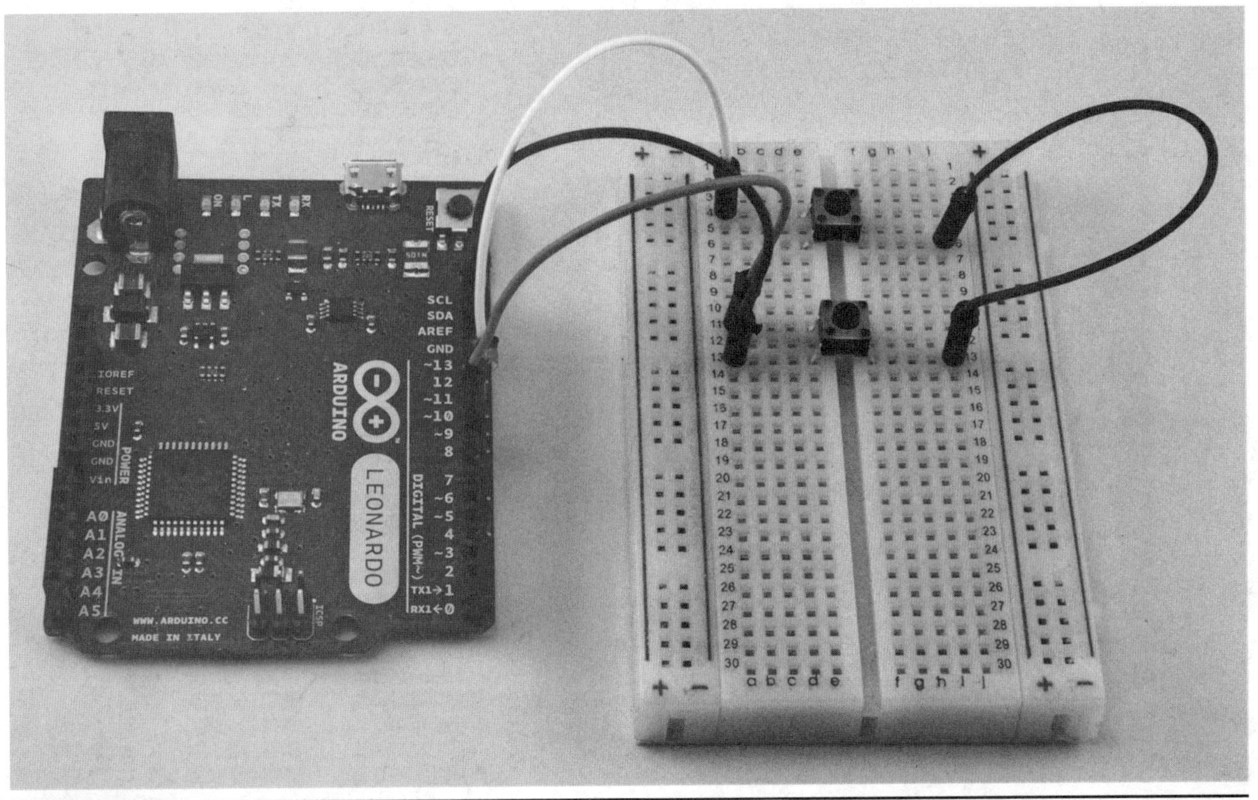

Figure 10-2 Automatic password typer.

Be warned: It is very easy to press the wrong button and accidentally reset the password. So think twice before using this project for your passwords. It is also not very secure because all someone needs to do to discover your password is to position the cursor in a word processor and then press the button for the password to be printed out in full view.

COMPONENTS AND EQUIPMENT

	Description	Appendix
	Arduino Leonardo	m2
S1, S2	Push to make switch	h3
	Solderless breadboard	h1
	Jumper wires	h2

Hardware

As far as hardware is concerned, this is one of the simplest projects in this book. It has just two push buttons attached to the breadboard.

Figure 10-3 shows the schematic diagram and Figure 10-4 the breadboard layout.

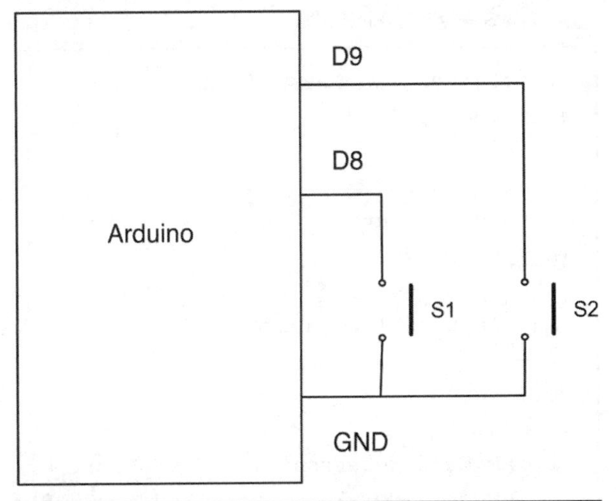

Figure 10-3 Schematic diagram for the password typer.

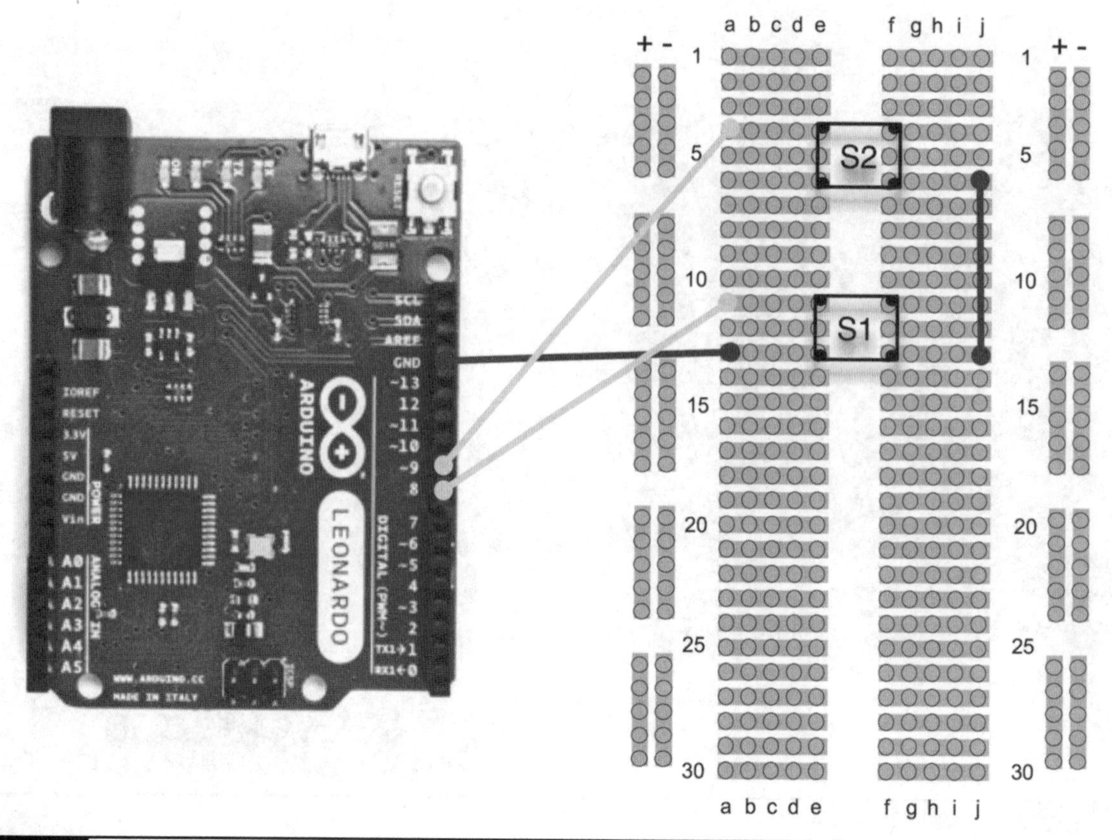

Figure 10-4 Breadboard layout for the password typer.

Software

The sketch for Project 32 is shown in Listing Project 32.

In addition to the two button variables, a variable passwordLength is also defined. If you want longer passwords, then all you need to do is increase this from 8 up to any value less than 1023. The character array (letters) is used to contain a list of the characters that can be used in the passwords that will be generated.

LISTING PROJECT 32

```
// Project 32 - Password Typer
#include <EEPROM.h>

int typeButton = 9;
int generateButton = 8;
int passwordLength = 8;

char letters[] = "abcdefghijklmnopqrstuvwxyzABCDEFGHIJKLMNOPQRSTUVWXYZ0123456789";

void setup()
{
  pinMode(typeButton, INPUT_PULLUP);
  pinMode(generateButton, INPUT_PULLUP);
  Keyboard.begin();
```

LISTING PROJECT 32 *(continued)*

```
}

void loop()
{
  if (digitalRead(typeButton) == LOW)
  {
    typePassword();
  }
  if (digitalRead(generateButton) == LOW)
  {
    generatePassword();
  }
  delay(300);
}

void typePassword()
{
  for (int i = 0; i < passwordLength; i++)
  {
    Keyboard.write(EEPROM.read(i));
  }
  Keyboard.write('\n');
}

void generatePassword()
{
  randomSeed(millis() * analogRead(A0));
  for (int i = 0; i < passwordLength; i++)
  {
    EEPROM.write(i, randomLetter());
  }
}

char randomLetter()
{
  int n = strlen(letters);
  int i = random(n);
  return letters[i];
}
```

For the Leonardo to do the clever keyboard trick, you need to use the command Keyboard.begin() in setup to start the keyboard emulation.

The loop function just needs to check for key presses and call either generatePassword or typePassword if the corresponding key is pressed.

The function typePassword simply reads each of the characters from EEPROM and echoes them to the keyboard using Keyboard.write. When all the letters have been written, it writes a final \n, which is the end-of-line character and simulates pressing the ENTER key.

To generate a new password, the pseudo-random-number generator is first seeded with a combination of the current milliseconds since last reboot and the value on the analog pin A0. Because both are pretty random, this will help to ensure that a nice random sequence of characters is created. These characters are created by repeatedly picking one of the characters from the array letters and writing it into EEPROM.

Putting It All Together

The best way to test out the project is to open something like Notepad on Windows or any kind of text editor. First, press the bottom button to generate a new password, and then press the top button, and the password should be typed for you into the editor (Figure 10-5).

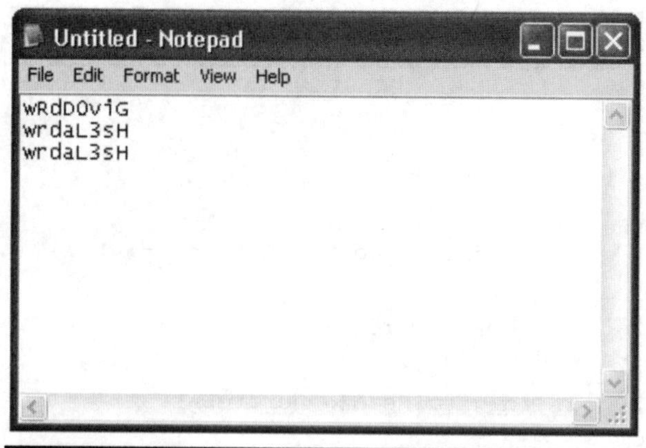

Figure 10-5 Using the password typer with Notepad on Windows.

Project 33
Accelerometer Mouse

This project turns a Leonardo into an accelerometer-controlled mouse with the help of a accelerometer module. Tip the Leonardo from side to side to control your mouse, and click on the push button to simulate a mouse click.

This project does not use breadboard; both the accelerometer module and the switch are attached directly to the Arduino header socket (Figure 10-6).

When the accelerometer is level, then the acceleration owing to gravity will act equally on both the X and Y dimensions. However, when you tip the module to one side, the amount of acceleration force changes in that dimension, and

Figure 10-6 Accelerometer mouse.

COMPONENTS AND EQUIPMENT	
Description	**Appendix**
Arduino Leonardo	m2
Push switch	h3
Adafruit acceleration module	m8

you can use these changes to send commands to alter the position of the mouse.

Hardware

Figure 10-7 shows the schematic diagram for the project.

The accelerometer module comes as a kit that includes the board and a short strip of headers that must be soldered to the connector. Follow the instructions for assembling this on the Adafruit website (http://www.adafruit.com/products/163). You do not need to solder the "Test" connection, but if you do, this pin can just hang off the right-hand side of the analog input sockets, as shown in Figure 10-6.

The push switch is also just connected on one side, between GND and D12.

Software

The sketch for Project 33 is shown in Listing Project 33.

LISTING PROJECT 33

```
// Project_33 Accelerometer Mouse

int gndPin = A2;
int xPin = 5;
int yPin = 4;
int zPin = 3;
int plusPin = A0;
int switchPin = 12;

void setup()
{
  pinMode(gndPin, OUTPUT);
  digitalWrite(gndPin, LOW);
  pinMode(plusPin, OUTPUT);
  digitalWrite(plusPin, HIGH);
  pinMode(switchPin, INPUT_PULLUP);
  pinMode(A1, INPUT); // 3V output
  Mouse.begin();
}

void loop()
{
  int x = analogRead(xPin) - 340;
  int y = analogRead(yPin) - 340;
  // midpoint 340, 340
  if (abs(x) > 10 || abs(y) > 10)
  {
    Mouse.move(x / 30, -y / 30, 0);
  }
  if (digitalRead(switchPin) == LOW)
  {
    Mouse.click();
    delay(100);
  }
}
```

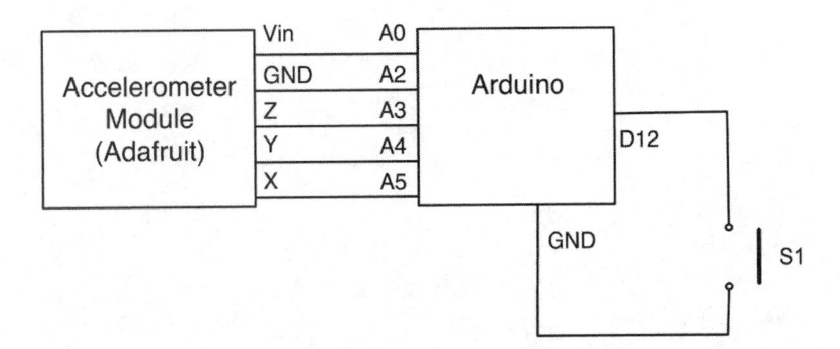

Figure 10-7 Schematic diagram for the accelerometer mouse.

The Leonardo uses three of the analog pins to measure the acceleration forces of the X, Y, and Z axes, but it also uses two of the analog pins (A2 and A0) to provide power to the accelerometer module. These are all set to their appropriate values in the setup function.

Pin A1 is set to be an input because the module actually outputs 3V on this pin, which we do not want to use, but A1 is one of the connections to which a header pin was attached. So setting it to be an input makes sure that it cannot conflict with the pin being used by the Arduino as an output and hence damaging the Arduino or module.

Making a Leonardo behave like a mouse is very similar to the previous two keyboard projects. You first have to start it impersonating a mouse by issuing the Mouse.begin command.

The loop function measures the acceleration forces of the X and Y axes, and then, if they are greater than the threshold of 10, it uses a scaled-down value of the acceleration offset to adjust the mouse position.

The switch is also checked in the loop, and if it is pressed, then the Mouse.click() command is sent.

Putting It All Together

Install the sketch for this project, and you should find that when you pick up your Leonardo, you will be able to control your mouse cursor by tilting it back and forth.

Pressing the button will perform the same action as clicking your regular mouse button.

Summary

The Leonardo is a very versatile device, and the projects in this chapter could be expanded in all sorts of directions. You could, for example, modify Project 32 to add lots of buttons and issue different keystrokes when the button was pressed to make a controller for musical software such as Ableton Live.

This is the final chapter containing projects. The author hopes that in trying the projects in this book, the Evil Genius' appetite for experimentation and design has been stirred, and he or she will have the urge to design some projects of his or her own.

Chapter 11 sets out to help you in the process of developing your own projects.

CHAPTER 11

Your Projects

So you have tried your hand at some of the author's projects and hopefully learned something along the way. Now it's time to start developing your own projects using what you have learned. You will be able to borrow bits of design from the projects in this book, but to help you along, this chapter gets you started with some design and construction techniques.

Circuits

The author likes to start a project with a vague notion of what he wants to achieve and then start designing from the perspective of the electronics. The software usually comes afterwards.

The way to express an electronic circuit is to use a schematic diagram. The author has included schematic diagrams for all the projects in this book, so even if you are not very familiar with electronics, you should now have seen enough schematics to understand roughly how they relate to the breadboard layout diagrams also included.

Schematic Diagrams

In a schematic diagram, connections between components are shown as lines. These connections will use the connective strips beneath the surface of the breadboard and the wires connecting one breadboard strip to another. For the kinds of projects in this book, it does not normally matter how the connection is made. The arrangement of the actual wires does not matter as long as all the points that should be connected are connected.

Schematic diagrams have a few conventions that are worth pointing out. For instance, it is common to place GND lines near the bottom of the diagram and higher voltages near the top of the diagram. This allows someone reading the schematic to visualize the flow of charge through the system from higher to lower voltages.

Another convention in schematic diagrams is to use the little bar symbol to indicate a connection to GND where there is not enough room to draw all the connections.

Figure 11-1, originally from Project 5, shows three resistors, all with one lead connected to the GND connection of the Arduino board. In the corresponding breadboard layout (Figure 11-2), you can see that the connections to GND go through three wires and three strips of breadboard connector block.

There are many different tools for drawing schematic diagrams. Some of them are integrated-electronics computer-aided design (CAD) products that will go on to lay out the tracks on a printed-circuit board for you. By and large, these create fairly ugly-looking diagrams, and the author prefers to use pencil and paper or general-purpose drawing software. All the diagrams for this book were created using Omni Group's excellent but strangely named OmniGraffle software, which is only available for Apple Macs. OmniGraffle

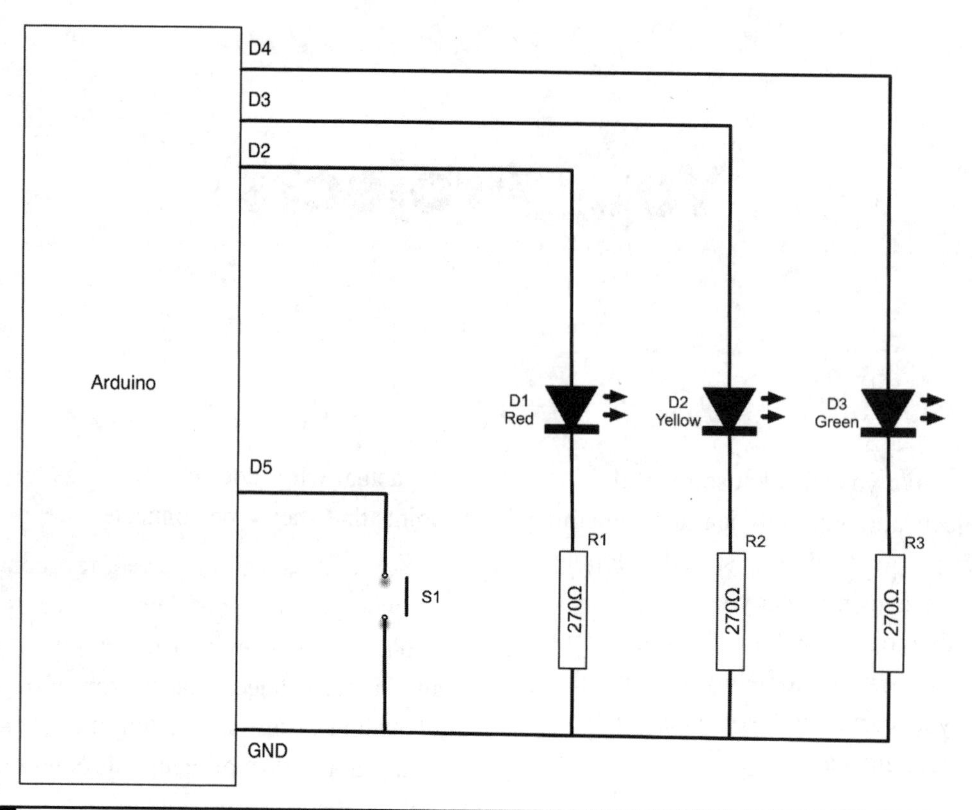

Figure 11-1 A schematic diagram example.

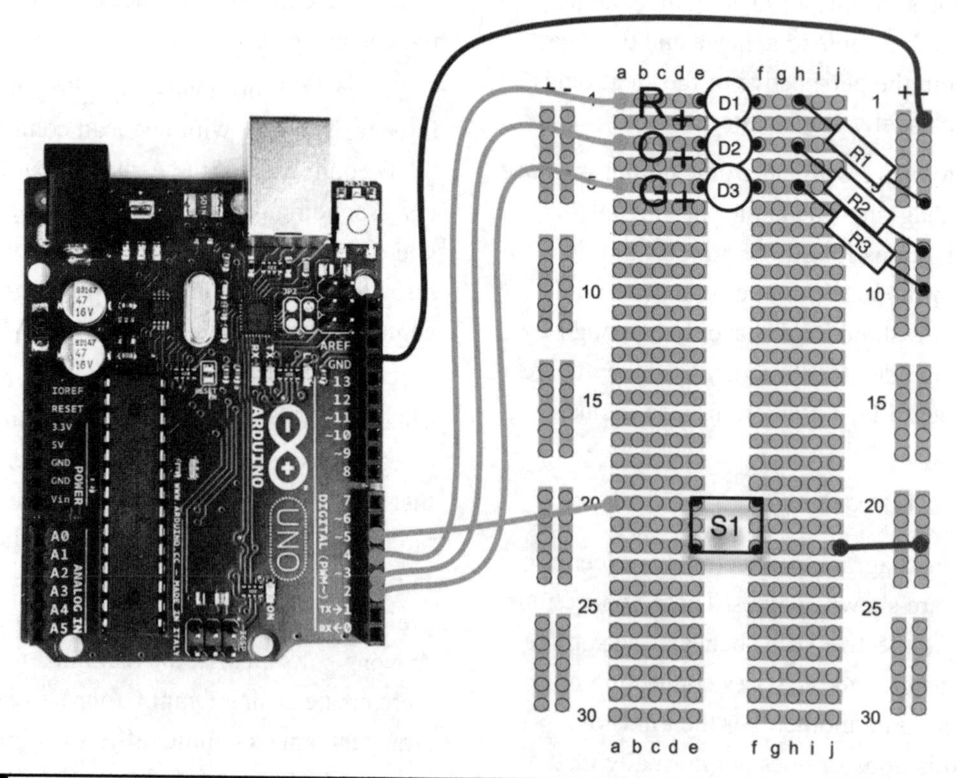

Figure 11-2 Example breadboard layout.

templates for drawing breadboard layouts and schematic diagrams are available for download from www.arduinoevilgenius.com.

Component Symbols

Figure 11-3 shows the circuit symbols for the electronic components that we have used in this book.

There are various different standards for circuit diagrams, but the basic symbols are all recognizable between standards. The set used in this book does not closely follow any particular standard. I have just chosen what I consider to be the most easy-to-read approach to the diagrams.

Components

In this section we look at the practical aspects of components: what they do and how to identify, choose, and use them.

Datasheets

All component manufacturers produce datasheets for their products. These act as a specification for how the component will behave. They are not of much interest for resistors and capacitors but are much more useful for semiconductors and transistors and especially integrated circuits. They will often include application notes that include example schematics for using the components.

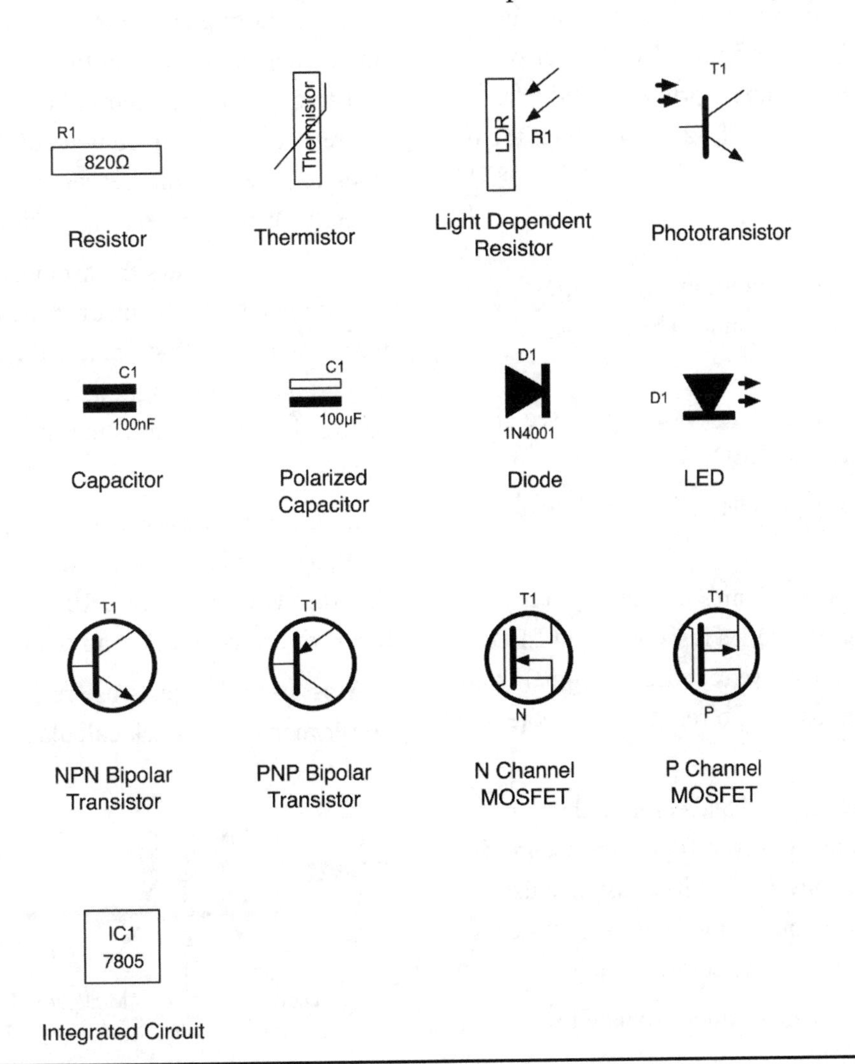

Figure 11-3 Circuit symbols.

These are all available on the Internet. However, if you search for "BC158 datasheet" in your favorite search engine, you will find that many of the top hits are for organizations cashing in on the fact that people search for datasheets a lot. These organizations surround the datasheets with pointless advertising and pretend that they add some value to looking up datasheets by subscribing to their service. These websites usually just lead to a frustration of clicking and should be ignored in favor of any manufacturers' websites. So scan through the search results until you see a URL such as www.fairchild.com.

Alternatively, many of the component retail suppliers such as Farnell provide free-of-charge datasheets for practically every component they sell, which is to be much applauded. This also means that you can compare prices and buy the components while you are finding out about them.

Resistors

Resistors are the most common and cheapest electronic components around. Their most common uses are

- To prevent excessive current flowing (see any project that uses an LED)
- In a pair or as a variable resistor to divide a voltage

Chapter 2 explained Ohm's law and used it to decide on a value of a series resistor for an LED. Similarly, in Project 19 we reduced the signal from our resistor ladder using two resistors as a potential divider.

Resistors have colored bands around them to indicate their value. However, if you are unsure of a resistor, you can always find its resistance using a multimeter. Once you get the hang of it, it's easy to read the values using the colored bands.

Each band color has a value associated with it, as shown in Table 11-1.

TABLE 11-1 Resistor Color Codes

Color	Value
Black	0
Brown	1
Red	2
Orange	3
Yellow	4
Green	5
Blue	6
Violet	7
Gray	8
White	9

There will generally be three of these bands together starting at one end of the resistor, a gap, and then a single band at the other end of the resistor. The single band indicates the accuracy of the resistor value. Since none of the projects in this book require accurate resistors, there is no need to select your resistors on this basis.

Figure 11-4 shows the arrangement of the colored bands. The resistor value uses just the three bands. The first band is the first digit, the second the second digit, and the third "multiplier" band is how many zeros to put after the first two digits.

So a 270 Ω resistor will have a first digit of 2 (red), a second digit of 7 (violet), and a multiplier of 1 (brown). Similarly, a 10 kΩ resistor will have bands of brown, black, and orange (1, 0, and 000).

Most of our projects use resistors in a very low-power manner. A quick calculation can be used to

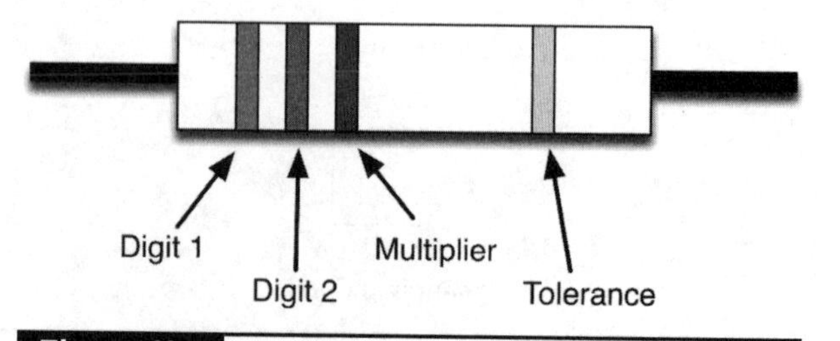

Figure 11-4 A color-coded resistor.

work out the current flowing through the resistor, and multiplying that number by the voltage across the resistor will tell you the power used by the resistor. The resistor burns off this surplus power as heat, so resistors will get warm if a significant amount of current flows through them.

You only need to worry about this for low-value resistors of less than 100 Ω or so because higher-value resistors will have such a small current flowing through them.

As an example, a 100 Ω resistor connected directly between 5V and GND will have a current through it of $I = V/R$, or 5/100, or 0.05 A. The power it uses will be $I \times V$, or $0.05 \times 5 = 0.25$ W.

A standard power rating for resistors is 0.5 or 0.6 W, and unless otherwise stated in projects, 0.5 W metal film resistors will be fine.

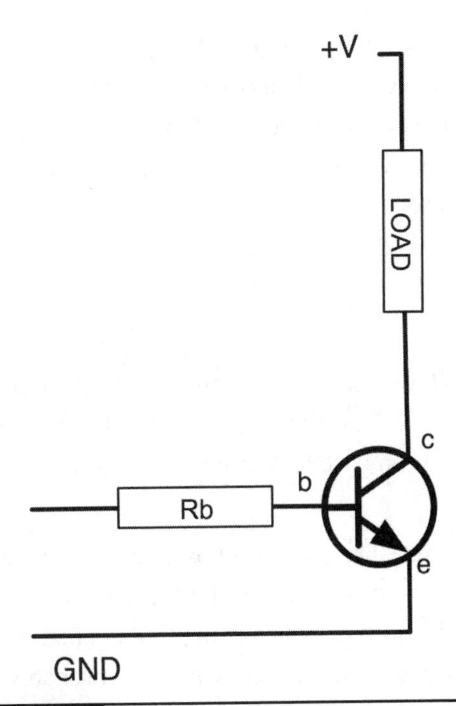

Figure 11-5 Basic transistor switch circuit.

Transistors

Browse through any component catalog and you will find literally thousands of different transistor types. In this book the list has been simplified to what's shown in Table 11-2.

The basic switch circuit for a transistor is shown in Figure 11-5.

The current flowing from base to emitter (b to e) controls the larger current flowing from the collector to the emitter. If no current flows into the base, then no current will flow through the load. In most transistors, if the load has zero resistance, the current flowing into the collector would be 50 to 200 times the base current. However, we are going to be switching our transistor fully on or fully off, so the load resistance will always limit the collector current to the current required by the load. Too much base current will damage the transistor and also rather defeat the objective of controlling a bigger current with a smaller one, so the base will have a resistor connected to it.

TABLE 11-2 Transistors Used in This Book

Transistor	Type	Purpose
2N2222	Bipolar NPN	Switching small loads greater than 40 mA.
BD139	Bipolar NPN power	Switching higher-load currents (e.g., Luxeon LED). See Project 6.
2N7000	N-channel FET	Low-power switching with very low "on" resistance. See Project 7.
FQP33N10	N-channel power MOSFET	High-power switching.
FQP27P06	P-channel power MOSFET	High-power switching.

When switching from an Arduino board, the maximum current of an output is 40 mA, so we could choose a resistor that allows about 30 mA to flow when the output pin is high at 5V. Using Ohm's law,

$$R = V/I$$

$$R = (5 - 0.6)/30 = 147$$

The –0.6 is because one characteristic of bipolar transistors is that there is always a voltage of about 0.6V between base and emitter when a transistor is turned on.

Therefore, using a 150 Ω base resistor, we could control a collector current of 40 to 200 times 30 mA, or 1.2 to 6 A, which is more than enough for most purposes. In practice, we would probably use a resistor of 1 kΩ or perhaps 270 Ω.

Transistors have a number of maximum parameter values that should not be exceeded or the transistor may be damaged. You can find these by looking at the datasheet for the transistor. For example, the datasheet for a 2N2222 will contain many values. The ones of most interest to us are summarized in Table 11-3.

TABLE 11-3 Transistor Datasheet

Property	Value	What It Means
I_c	800 mA	The maximum current that can flow through the collector without the transistor being damaged.
h_{FE}	100–300 mA	DC current gain. This is the ratio of collector current to base current, and as you can see, it could be anything between 100 and 300 mA for this transistor.

Other Semiconductors

The various projects have introduced a number of different types of components, from LEDs to temperature sensors; Table 11-4 provides some pointers into the various projects. If you want to develop your own project that senses temperature or whatever, first read about the projects developed by the author that use these components.

TABLE 10-4 Use of Specialized Components in Projects

Component	Project
Single-color LEDs	Almost every project
Multicolor LEDs	14
LED matrix displays	16
7-segment LEDs	15, 30
Audio amplifier chip	19, 20
LDR (light sensor)	20
Variable voltage regulator	7

It may even be worth building the project and then modifying it to your own purposes.

Modules and Shields

It does not always make sense to make everything from scratch. This is why, after all, we buy an Arduino board rather than make our own. The same is true of some modules that we may want to use in our projects.

For instance, the LCD display module that we used in Projects 17 and 22 contains the driver chip needed to work the LCD itself, reducing both the amount of work we need to do in the sketch and the number of pins we need to use.

Other types of modules are available that you may wish to use in your projects. Suppliers such as Sparkfun and Adafruit are a great source of ideas and modules. A sample of the kinds of modules that you can get from such suppliers includes

- GPS
- Wi-Fi
- Bluetooth
- Zigbee wireless
- GPRS cellular modem

You will need to spend time reading through datasheets, planning, and experimenting, but that is what being an Evil Genius is all about.

Slightly less challenging than using a module from scratch is to buy an Arduino shield with the module already installed. This is a good idea when the components that you would like to use will not go on a breadboard (such as surface-mount devices). A ready-made shield can give you a real leg up with a project.

New shields become available all the time, but at the time of this writing, you can buy Arduino shields for

- Ethernet (connect your Arduino to the Internet)
- XBee (a wireless data-connection standard used in home automation, among other things)
- Motor driver
- GPS
- Joystick
- SD card interface
- Graphic LCD touch-screen display
- Wi-Fi

Buying Components

Thirty years ago, the electronics enthusiast living in even a small town would be likely to have the choice of several radio/TV repair and spare stores where he or she could buy components and receive friendly advice. These days there are a few retail outlets that still sell components, such as RadioShack in the United States and Maplins in the United Kingdom, but the Internet has stepped in to fill the gap, and it is now easier and cheaper than ever to buy components.

With component suppliers such as Digikey, Mouser, Newark, Radio Spares, and Farnell, you can fill a virtual shopping basket online and have the components arrive in a day or two. Shop around because prices vary considerably among suppliers for the same components.

You will find eBay to be a great source of components. If you don't mind waiting a few weeks for your components to arrive, there are great bargains to be had from China. You often have to buy large quantities but may find it cheaper to get 50 of a component from China than 5 locally. In this way, you have some spares for your component box.

Tools

When making your own projects, there are a few tools that you will need at a bare minimum. If you do not intend to do any soldering, then you will need

- Solid-core wire in a few different colors, something around 0.6 mm (23 SWG) diameter
- Pliers and wire snips, particularly for making jumper wires for the breadboard
- Breadboard
- Multimeter

If you intend to solder, then you will also need

- Soldering iron (duh)
- Lead-free alloy solder

Component Box

When you first start designing your own projects, it will take you some time to gradually build up your stock of components. Each time you are finished with a project, a few more components will find their way back to your stock.

It is useful to have a basic stock of components so that you do not have to keep ordering things when you just need a different-value resistor. You will have noticed that most of the projects in this book tend to use resistor values such as 100 Ω, 1 kΩ, 10 kΩ, etc. You actually don't need that many different components to cover most of the bases for a new project.

A good starting kit of components is listed in the Appendix.

Boxes with compartments that can be labeled save a lot of time in selecting components, especially resistors that do not have their value written on them.

Snips and Pliers

Snips are for cutting, and pliers are for holding things still (often while you cut them).

Figure 11-6 shows how you strip the insulation off wire. Assuming that you are right-handed, hold your pliers in your left hand and the snips in the right. Grip the wire with the pliers close to where you want to start stripping the wire, and then gently pinch round the wire with the snips and pull sideways to pull the insulation away. Sometimes you will pinch too hard and cut or weaken the wire, and other times you will not pinch hard enough and the insulation will remain intact. It's all just a matter of practice.

You also can buy an automatic wire stripper that grips and removes insulation in one action. In practice, these often only work well for one particular wire type and sometimes just plain don't work.

Soldering

You do not have to spend a lot of money to get a decent soldering iron. Temperature-controlled solder stations, such as the one shown in Figure 11-7, are better, but a fixed-temperature electric iron is fine. Buy one with a fine tip, and make sure that it is intended for electronics and not plumbing use.

Use narrow lead-free solder. Anyone can solder things together and make them work, but some people just have a talent for neat soldering. Don't worry if your results do not look as neat as a robot-made printed circuit. They are never going to.

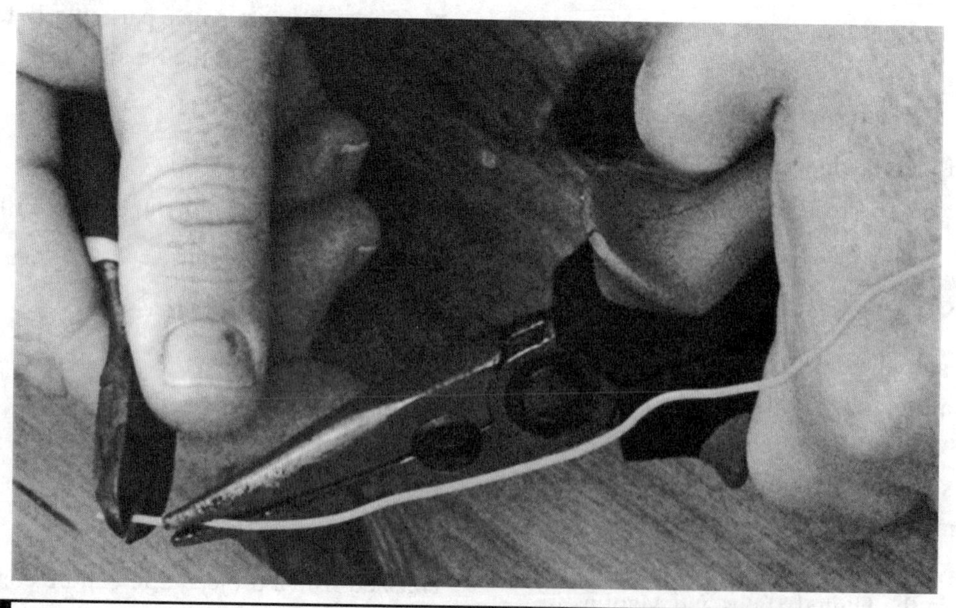

Figure 11-6 Snips and pliers.

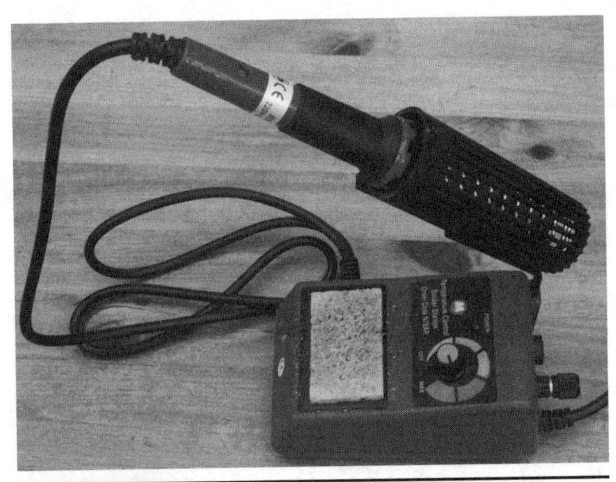

Figure 11-7 Soldering iron and solder.

Soldering is one of those jobs that you really need three hands for: one hand to hold the soldering iron, one to hold the solder, and one to hold the thing you are soldering. Sometimes the thing you are soldering is big and heavy enough to stay put while you solder it; on other occasions, you will need to hold it down. Heavy pliers are good for this, as are minivises and "helping hand" type holders that use little clips to grip things.

The basic steps for soldering are

1. Wet the sponge in the soldering iron stand.
2. Allow the iron to come up to temperature.
3. Tin the tip of the iron by pressing the solder against it until it melts and covers the tip.
4. Wipe the tip on the wet sponge—this produces a satisfying sizzling sound but also cleans off the excess solder. You should now have a nice bright silver tip.
5. Touch the iron to the place where you are going to solder to heat it; then after a short pause (a second or two), touch the solder to the point where the tip of the iron meets the thing you are soldering. The solder should flow like a liquid, neatly making a joint.
6. Remove the solder and soldering iron, putting the iron back in its stand and being very careful that nothing moves in the few seconds that the solder will take to solidify. If something does move, then touch the iron to it again to reflow the solder; otherwise, you can get a bad connection called a *dry-joint*.

Above all, try not to heat sensitive (or expensive) components any longer than necessary, especially if they have short leads.

Practice soldering any old bits of wire together or wires to an old bit of circuit board before working on the real thing.

Multimeter

A big problem with electrons is that you cannot see the little monkeys. A multimeter allows you to see what they are up to. It allows you to measure voltage, current, resistance, and often other features too, such as capacitance and frequency. A cheap $10 multimeter is perfectly adequate for almost any purpose. The professionals use much more solid and accurate meters, but they're not necessary for most purposes.

Multimeters, such as the one shown in Figure 11-8, can be either analog or digital. You can tell more from an analog meter than you can from a digital meter because you can see how fast a needle swings over and how it jitters, something

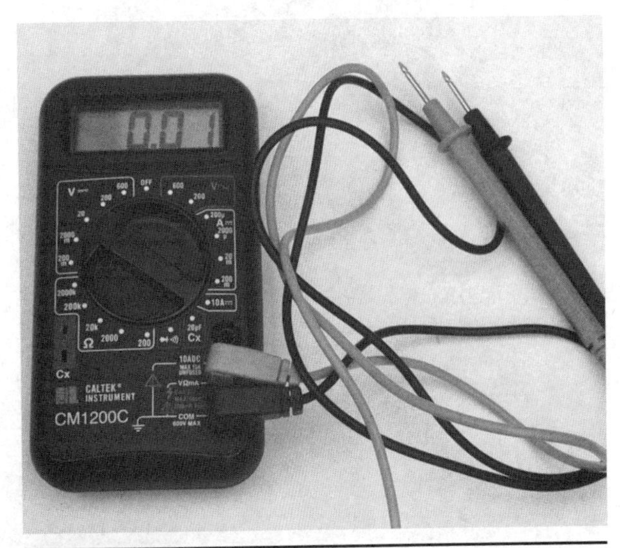

Figure 11-8 A multimeter.

that is not possible with a digital meter, where the numbers just change. However, for a steady voltage, it is much easier to read a digital meter because an analog meter will have a number of scales, and you have to work out which scale you should be looking at before you take the reading.

You can also get autoranging meters, which, once you have selected whether you are measuring current or voltage, will automatically change ranges for you as the voltage or current increases. This is useful, but some would argue that thinking about the range of voltage before you measure it is actually a useful step.

To measure voltage using a multimeter:

1. Set the multimeter range to voltage (start at a range that you know will be higher than the voltage you are about to measure).
2. Connect the black lead to GND. A crocodile clip on the negative lead makes this easier.
3. Touch the red lead to the point whose voltage you want to measure. For instance, to see if an Arduino digital output is on or off, you can touch the red lead to the pin and read the voltage, which should be either 5V or 0V.

Measuring current is different from measuring voltage because you want to measure the current flowing through something and not the voltage at some point. So you put the multimeter in the path of the current that you are measuring. This means that when the multimeter is set to a current setting, there will be a very low resistance between the two leads, so be careful not to short anything out with the leads.

Figure 11-9 shows how you could measure the current flowing through an LED.

To measure current:

1. Set the multimeter range to a current range higher than the expected current. Note that multimeters often have a separate high-current connector for currents as high as 10 A.

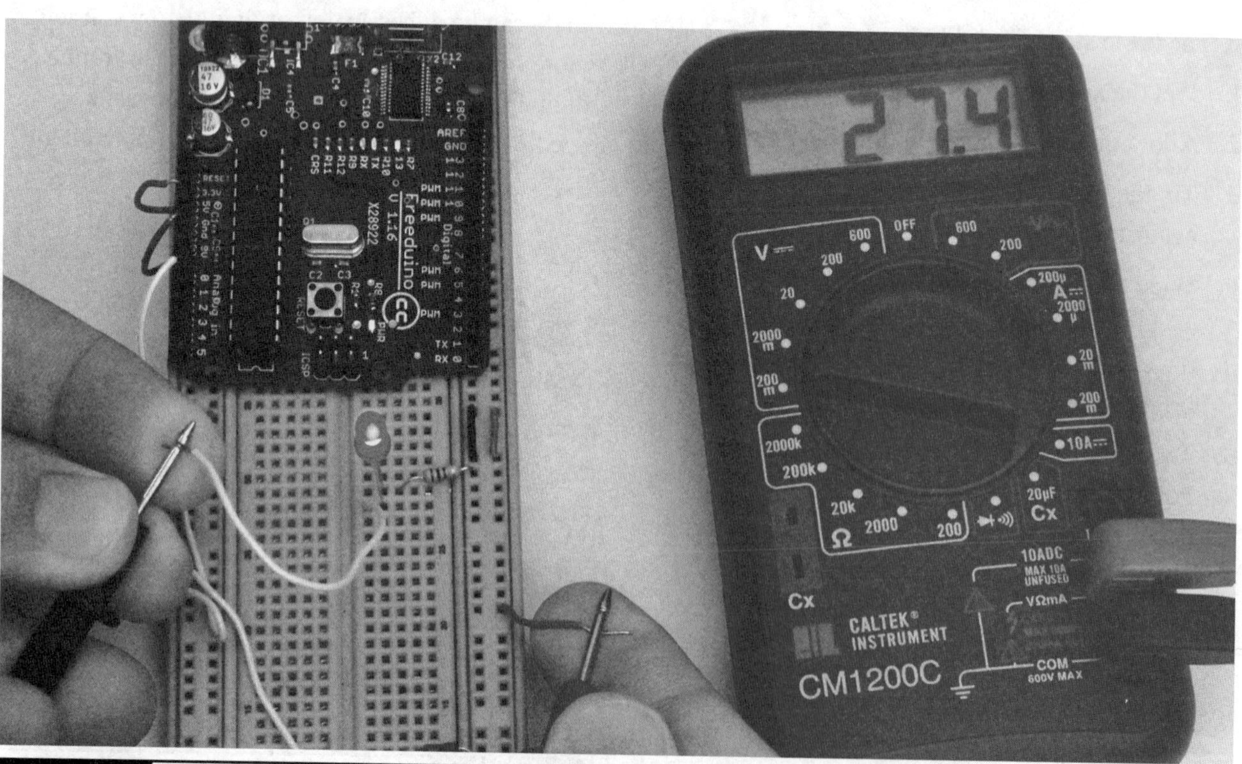

Figure 11-9 Measuring current.

2. Connect the positive lead of the meter to the more positive side from which the current will flow.
3. Connect the negative lead of the meter to the more negative side. Note that if you get this the wrong way round, a digital meter will just indicate a negative current; however, connecting an analog meter the wrong way round may damage it.
4. In the case of an LED, the LED should still light as brightly as before you put the meter into the circuit, and you will be able to read the current consumption.

Another feature of a multimeter that is sometimes useful is the continuity test feature. This will usually beep when the two test leads are connected together. You can use this to test fuses, etc., as well as to test for accidental short circuits on a circuit board or broken connections in a wire.

Resistance measurement is occasionally useful, particularly if you want to determine the resistance of an unmarked resistor.

Some meters also have diode and transistor test connections, which can be useful to find and discard transistors that have burned out.

Oscilloscope

In Project 18 we built a simple oscilloscope. An oscilloscope is an indispensable tool for any kind of electronics design or test where you are looking at a signal that changes over time. Oscilloscopes are relatively expensive, and there are various types. One of the most cost-effective types is similar in concept to the one in Project 18. That oscilloscope just sends its readings across to a computer that is responsible for displaying them.

Entire books have been written about using an oscilloscope effectively, and every oscilloscope is different, so we will just cover the basics here.

As you can see from Figure 11-10, the screen showing the waveform is displayed over the top of a grid. The vertical grid is in units of some fraction of volts, which on this screen is 2V per division. So the voltage of the square wave in total is 2.5 × 2 = 5V.

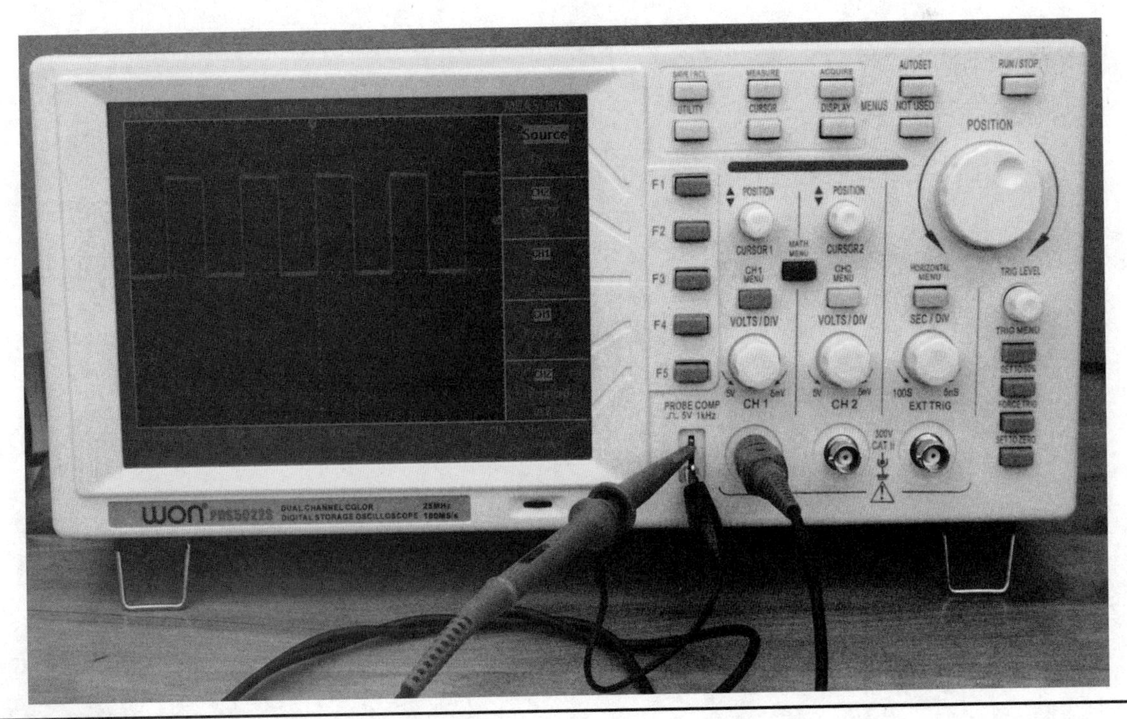

Figure 11-10 An oscilloscope.

The horizontal axis is the time axis, and this is calibrated in seconds—in this case, 500 ms (microseconds) per division. So the length of one complete cycle of the wave is 1000 ms, that is, 1 ms (millisecond), indicating a frequency of 1 kHz.

Project Ideas

The Arduino Playground on the main Arduino website (www.arduino.cc) is a great source of ideas for projects. Indeed, it even has a section specifically for project ideas, divided into easy, medium, or difficult.

If you type "Arduino project" into your favorite search engine or YouTube, you will find no end of interesting projects that people have embarked on.

Another source of inspiration is the component catalog, either online or on paper. Browsing through, you might come across an interesting component and wonder what you could do with it. Thinking up a project is something that should be allowed to gestate in the mind of the Evil Genius. After exploring all the options and mulling everything over, the Evil Genius' project will start to take shape!

If you enjoyed reading this book, you might like to consider some of the author's other books on Arduino and other areas of electronics. Please see www.simonmonk.org for a full list.

APPENDIX

Components and Supplies

ALL OF THE PARTS USED in this book are readily available through the Internet. However, sometimes it is a little difficult to track down exactly what you are looking for. For this reason, this appendix lists the components along with some order codes for various suppliers.

Suppliers

There are so many component suppliers out there that it feels a little unfair to list the few that the author knows. So have a look around on the Internet, as prices vary considerably between suppliers.

Some smaller suppliers specialize in providing components for home constructors building microcontroller projects like ours. They do not have the range of components, but do often have more exotic and fun components at reasonable prices. Great examples of this kind of supplier are Adafruit and Sparkfun Electronics, but there are many others out there.

Sometimes, when you find you just need a couple of components, it's great to be able to go to a local store and pick them up. RadioShack in the United States and Maplins in the UK stock a range of components, and are great for this purpose.

CPC (cpc.farnell.com) in the United Kingdom also sells a lot of Arduino related kit and bulk components such as resistors and capacitors at low cost.

Buying components can be quite daunting and buying something like Adafruit's Arduino experimenter kit (product ID 170) or Sparkfun Arduino Inventor's kit (KIT-11227) is a good way to get started with a basic selection of components and some breadboard.

The sections that follow list components by type, along with some possible sources and order codes where available.

Component Sources

The Component boxes for each project list Appendix codes for the components use. This section lists those codes and offers some sources from which they can be obtained.

The components are grouped into sections, each section being prefixed with a letter, M for module, R for resistor, etc.

Arduino and Module		
Code	**Description**	**RS**
m1	Arduino Uno R3	Adafruit: 50 Sparkfun: DEV-11021
m2	Arduino Leonardo	Adafruit: 849 Sparkfun: DEV-11286
m3	Arduino Lilypad	Sparkfun: DEV-09266
m4	Protoshield Kit	eBay
m5	8 by 8 LED bicolor I2C module	Adafruit: 902
m6	LCD Module (HD44780 controller)	Adafruit: 181 Sparkfun: LCD-00255
m7	I²C 4 digit 7 segment display	Adafruit: 880
m8	Adafruit Acceleration Module	Adafruit: 163

Resistors

Resistors are low-cost components, and you will often find that suppliers will only sell you them in quantities of 50 or 100. For common values like 270 Ω, 1 kΩ and 10 kΩ it can be really useful to have a bit of stock.

You can also buy resistor kits that have a wide range of resistors in a book or component box. If the kit of resistors does not have exactly the right value, then using the next value up will usually be fine. So, for instance, this book uses a lot of 270 Ω resistors with LEDs, but if your kit did not have this value, then using 300 Ω instead would work just fine.

Some resistor kits to check out are:

- Sparkfun: COM-10969
- Maplins: FA08J

Resistors		
Code	**Description**	**Sources**
r1	4.7 Ω 1/4W resistor	Digikey: S4.7HCT-ND Mouser: 293-4.7-RC CPC: RE06232
r2	100 Ω 1/4W resistor	Digikey: S100HCT-ND Mouser: 293-100-RC CPC: RE03721
r3	270 Ω 1/4W resistor	Digikey: 293-100-RC Mouser: 293-100-RC CPC: RE03747
r4	470 Ω 1/4W resistor	Digikey: 293-470-RC Mouser: 293-470-RC CPC: RE03799

Resistors (*continued*)		
Code	**Description**	**Sources**
r5	1 kΩ 1/4W resistor	Digikey: S1kHCT-ND Mouser: 293-1K-RC CPC: RE03722
r6	10 kΩ 1/4W resistor	Digikey: S10KHCT-ND Mouser: 293-10K-RC CPC: RE03723
r7	56 kΩ 1/4W resistor	Digikey: S56KHCT-ND Mouser: 273-56K-RC CPC: RE03764
r8	100 kΩ 1/4W resistor	Digikey: S100KHCT-ND Mouser: 273-100K-RC CPC: RE03724
r9	470 kΩ 1/4W resistor	Digikey: S470KHCT-ND Mouser: 273-470K-RC CPC: RE0375
r10	1 MΩ 1/4W resistor	Digikey: S1MHCT-ND Mouser: 293-1M-RC CPC: RE03725
r11	10 kΩ linear potentiometer (trimpot)	Adafruit: 356 Sparkfun: COM-09806 Digikey: 3362P-103LF-ND Mouser: 652-3362P-1-103LF CPC: RE06517
r12	100 kΩ linear potentiometer (full size)	Digikey: 987-1312-ND Mouser: 858-P120KGPF20BR100K CPC: RE04393
r13	LDR	Adafruit:161 Sparkfun: SEN-09088 Digikey: PDV-P8001-ND CPC: RE00180
r14	10 Ω 1/2W resistor	Digikey: S10HCT-ND Mouser: 293-10-RC CPC: RE05005

Capacitors		
Code	Description	Sources
c1	100 nF	Adafruit: 753 Sparkfun: COM-08375 Digikey: 445-5258-ND Mouser: 810-FK18X7R1E104K CPC: CA05514
c2	220 nF	Digikey: 445-2849-ND Mouser: 810-FK16X7R2A224K CPC: CA05521
c3	100 F electrolytic	Sparkfun: COM-00096 Digikey: P5529-ND Mouser: 647-UST1C101MDD CPC: CA07510

Semiconductors

This book uses a lot of LEDs, so it is worth looking around for an LED kit, rather than buying the size and color combinations separately. As well as very cheap LED selections available direct from China, Maplins and other suppliers sell starter kit of assorted LEDs (product code RS37S).

Semiconductors		
Code	Description	Sources
s1	5mm red LED	Adafruit: 297 Sparkfun: COM-09590 Digikey: 751-1118-ND Mouser: 941-C503BRANCY0B0AA1 CPC: SC11574
s2	5mm green LED	Adafruit: 298 Sparkfun: COM-09650 Digikey: 365-1186-ND Mouser: 941-C503TGANCA0E0792 CPC: SC11573
s3	5mm yellow LED	Sparkfun: COM-09594$0.35 Digikey: 365-1190-ND Mouser: 941-C5SMFAJSCT0U0342 CPC: SC11577
s4	2 or 3 mm red LED	Sparkfun: COM-00533 Digikey: 751-1129-ND Mouser: 755-SLR343BCT3F CPC: SC11532

Semiconductors (*continued*)		
Code	**Description**	**Sources**
s5	2 or 3 mm green LED	Sparkfun: COM-09650 Digikey: 751-1101-ND Mouser: 755-SLR-342MG3F CPC: SC11533
s6	2 or 3 mm blue LED	Digikey: 751-1092-ND Mouser: 755-SLR343BC7T3F CPC: SC11560
s7	RGB LED common cathode	Sparkfun: COM-09264
s8	Two-digit, seven-segment LED display (common anode)	Mouser: 604-DA03-11YWA
s9	10 segment bar graph display	Farnell: 1020492 CPC: SC12044
s10	Luxeon 1 W LED	Adafruit: 518 Sparkfun: BOB-09656 Digikey: 160-1751-ND Mouser: 859-LOPL-E011WA CPC: SC11807
s11	3 mW red laser diode module	eBay
s12	1N4004 or 1N4001 diode	Adafruit: 755 Sparkfun: COM-08589 Digikey: 1N4001-E3/54GITR-ND Mouser: 512-1N4001 CPC: SC07332
s13	5.1V zener diode	Sparkfun: COM-10301 Digikey: 1N4733AVSTR-ND Mouser: 1N4733AVSTR-ND CPC: SC07166
s14	2N2222 or BC548 or 2N3904 NPN transistor	Sparkfun: COM-00521 Digikey: 2N3904-APTB-ND Mouser: 610-2N3904 CPC: SC12549
s15	2N7000 FET	Digikey: 2N7000TACT-ND Mouser: 512-2N7000 CPC: SC06951
s16	FQP30N06 transistor	Adafruit: 355 Sparkfun: COM-10213 Digikey: FQP30N06L-ND Mouser: 512-FQP30N06 CPC: SC08210
s17	BD139 power transistor	Digikey: BD13916STU-ND Mouser: 511-BD139 CPC: SC09455

(continued on next page)

Semiconductors (*continued*)		
Code	**Description**	**Sources**
s18	LM317 voltage regulator	Digikey: 296-13869-5-ND Mouser: 595-LM317KCSE3 CPC: SC08256
s19	IR phototransistor 940 nm	Digikey: 365-1067-ND Mouser: 828-OP505B CPC: SC08558
s20	5mm IR LED sender 940 nm	Digikey: 751-1203-ND Mouser: 782-VSLB3940 CPC: SC1236
s21	IR remote control receiver IC	Mouser: 782-TSOP4138 CPC: SC12388
s22	TMP36 temperature sensor	Adafruit: 165 Sparkfun: SEN-10988 Digikey: TMP36GT9Z-ND CPC: SC10437
s23	TDA7052 1W audio amplifier	Digikey: 568-1138-5-ND Mouser: 771-TDA7052AN CPC: SC08454
s24	L293D motor driver	Adafruit: 807 Sparkfun: COM-00315 Digikey: 296-9518-5-ND Mouser: 511-L293D CPC: SC10241

Hardware and Miscellaneous

Most of the items in this section will be available on eBay at a low cost.

Hardware and Miscellaneous		
Code	**Description**	**Sources**
h1	Solderless breadboard	Adafruit: 64 Sparkfun: PRT-09567
h2	Set of jumper wires	Adafruit: 758 Sparkfun: PRT-08431
h3	Miniature push to make switch	Adafruit: 1119 Sparkfun: COM-00097 Digikey: SW853-ND Mouser: 653-B3W-1100

Hardware and Miscellaneous *(continued)*		
Code	**Description**	**Sources**
h4	2.1mm DC power jack	Digikey: SC1052-ND Mouser: 502-S-760 CPC: CN14795
h5	9V battery clip	Digikey: BS61KIT-ND Mouser: 563-HH-3449 CPC: BT03732
h6	Regulated 5V 1A power supply	Most suppliers or eBay. Country-specific connectors.
h7	Regulated 12V 2A power supply	
h8	Regulated 15V 1A power supply	
h9	Perf board	Farnell: 1172145 CPC: PC01222
h10	Three-way screw terminal	Farnell: 1641933
h11	4 by 3 keypad	Adafruit: 419 Sparkfun: COM-08653
h12	0.1-inch header strip	Adafruit: 392
h13	Rotary encoder with push switch	Digikey: CT3011-ND Mouser: 774-290VAA5F201B2 Farnell: 1520815
h14	Miniature 8 Ω loudspeaker	Sparkfun: COM-09151 Farnell: 1300022
h15	Electret microphone	Sparkfun: COM-08635 Digikey: 102-1721-ND Mouser: 665-POM2738PC33R Farnell: 1736563
h16	5V relay	Digikey: T7CV1D-05-ND Mouser: 893-833H-1C-S-5VDC CPC: SW03694
h17	12V cooling fan	eBay
h18	6V DC gearmotor	eBay
h19	Wheel to fit gearmotor	eBay
h20	9g servo motor	eBay Sparkfun: ROB-09065 Adafruit: 169
h21	Piezo buzzer	Adafruit: 160 Sparkfun: COM-07950
h22	Miniature reed switch	Sparkfun: COM-08642 Farnell: 1435590 CPC: SW00759
h23	Magnetic door latch	Farnell: COM-08642 CPC: SR04745

Index

References to figures are in italics.

U

V

W